7M/3C

BIOGRAPHY OF RESISTANCE

The Epic Battle Between People and Pathogens

BIOGRAPHY

OF

RESISTANCE

MUHAMMAD H. ZAMAN, PhD

HARPER WAVE

An Imprint of HarperCollins*Publishers*

HarperCollins books may be purchased for educational, business, or sales promotional use. For information, please email the Special Markets Department at SPsales@harpercollins.com.

FIRST EDITION

Library of Congress Cataloging-in-Publication Data has been applied for.

ISBN 978-0-06-286297-6

20 21 22 23 24   LSC   10 9 8 7 6 5 4 3 2 1

To Ammi and Abbu

CONTENTS

BIOGRAPHY OF RESISTANCE

PROLOGUE

Washoe County is on the western edge of Nevada, with Oregon to the north and California to the west. The county, with its picturesque lakes and stunning deserts, is not often in the news. But on January 13, 2017, a brief article from Washoe's public health officials was published in the Centers for Disease Control's *Mortality and Morbidity Weekly Report,*[1] and it sent shock waves around the world. It was the first report of its kind—never before had a US county public health office written about a complete failure of every single antibacterial drug that they had available to them.

The report concerned a Washoe County resident in her seventies who had been admitted to a hospital in Reno about five months earlier, showing signs of inflammation and infection. She had recently returned from an extended trip to India, where she had fallen and broken her femur, the largest bone in the human body.[2] She had been treated at a local hospital, and her condition improved, but later she developed an infection in her femur and hip. She was in and out of Indian hospitals as doctors tried to help her.

In August 2016, the doctors in Reno examined the woman and sent her blood and urine samples to the lab. The tests came back with results indicating that the bacteria causing her infection was resistant to leading antibiotics. The bacteria in question was CRE: carbapenem-resistant Enterobacteriaceae.[3] Enterobacteriaceae is a large family of bacteria, many of which live harmlessly in the human gut—but there are other types that are notoriously difficult to treat because they are resistant to highly potent antibiotics. In the case of this patient, the subset of bacteria belonging to the Enterobacteriaceae family was *Klebsiella pneumoniae*, which can cause pneumonia or sepsis and is a major cause of urinary tract infections.[4] And it can be life-threatening as well.

The doctors in Reno found this highly unusual. They had never seen evidence of CRE in their wards. Concerned that the very serious infection could easily spread to the other patients, the staff moved the woman to an acute care ward. The nurses and doctors attending the patient adopted the most stringent protocols for infection control, putting on gloves, masks, and extra layers of gowns any time they came in contact with her.

Resistance to antibiotics is a problem that infectious disease doctors encounter often in their line of work. If one drug isn't effective, they try others, and sometimes they even combine a few drugs to overcome a particularly difficult bacterial infection. Knowing that the most common antibiotics wouldn't help in this case, the doctors went to the next line of more potent drugs.

Most of the time doctors in the United States and other developed countries can find something that works. The treatment can be taxing on the patient, and recovery can be prolonged, but not everyone who gets CRE ends up dying.[5] Many recover fully, their lives saved by doctors eventually hitting upon the right drug, or combination of drugs, that kills the infection. Hoping that something might work, as is often the case, the doctors in Reno kept trying, determined to find an antibiotic that would save their patient. But this time was different—one antibiotic after another failed, combination after combination failed. Nothing seemed to work. The infection spread throughout her bloodstream and her organs. They used every antibiotic that was available in the United States at the time—a total of twenty-six. The infection thrived. The woman died of septic shock two weeks after arriving at the hospital in Reno.

Meanwhile, doctors on the other side of the world were facing an unprecedented challenge as well. Karachi, the largest city in my home country of Pakistan, could not be more different from Washoe County—it's a port city, a sprawling urban metropolis of nearly 15 million people, and among the most densely populated parts of the world.

In the fall of 2016, a typhoid outbreak in and around Karachi was proving extraordinarily difficult to control.[6] The typhoid was resistant to most frontline drugs. It was caused by another member of the Enterobacteriacae family that the doctors encountered in Nevada—*Salmonella typhi*. The outbreak that started in 2016 would last for nearly four years and affect thousands of not just people in Pakistan but people who traveled to and from the country as well.

For Karachi's citizens, this outbreak of a resistant bug was unprecedented. Typhoid is not uncommon in Pakistan, but a number of readily available antibiotics were proving no longer effective. Ultimately, there were only two left—carbapenem antibiotics and azithromycin.[7] Carbapenems are expensive, must be given intravenously, and require hospitalization—something many Pakistanis living in Karachi cannot afford. For them, their lifeline depended on the efficacy of the other option—azithromycin. Doctors and public health experts worried about the day when that option would no longer work.

Their fear is not unfounded. Bacteria mutate quickly and can also acquire resistance from other members of their family. What if the next outbreak is resistant to azithromycin as well? And what happens if that outbreak spreads, not just across a city as large as Karachi but throughout the country, or around the globe?

The Pakistan outbreak was classified as XDR—extensively drug resistant, the worst-case scenario. Worried about the threat, the CDC issued a warning for people traveling to Pakistan. Still, at least six people who had recently traveled to Pakistan during the outbreak came back to the United States and were diagnosed with XDR typhoid.[8] Canada and the United Kingdom also saw patients with XDR typhoid—all had traveled to Pakistan. Drugs available in Canada and the United States proved effective, and all of the patients survived, but many in Pakistan did not.

Patients across age groups, geographies, and economies are connected in this web of untreatable infections. It is not just a problem of resources or poverty, as some of the most advanced health-care systems in the world are struggling to manage drug-resistant infections. In the United States alone, well over thirty-five thousand people die every year due to multi-drug-resistant infections[9]—some of them in highly reputable hospitals. More people around the world die due to drug-resistant infections than breast cancer or HIV/AIDS or complications due to diabetes. While cancer and HIV/AIDS deaths are declining in the United States, and in many parts of the world, deaths due to drug resistance are constantly, rapidly increasing.

Connected across continents, countries, and cultures, antibiotic resistance is a danger to all of us. James Johnson, a prominent infectious disease specialist and expert in antimicrobial resistance in the United States, put it well when a journalist asked him how close we were to falling off the cliff into a world where our antibiotics no longer work. His response was simple: "We are already off the cliff."[10]

Similar warnings and declarations have been made before. And yet somehow, through discoveries during war and in times of peace, through genius and serendipity, in pursuit of profits and in demonstrations of compassion, scientists have been able to delay a total apocalypse. But is this time different? Are we near the end of our luck in the battle of people and pathogens? How much time do we have left?

CHAPTER 1

WHAT WE'RE UP AGAINST

Bacteria have been around far longer than humans—about 3.5 billion years longer—and they also outnumber us, by a lot. There are more bacteria on Earth than there are stars in the universe, and there are about 40 trillion in the human body alone.[1] Bacteria live in environments that are considered too harsh for any other form of life to exist—some live in the hot springs of Yellowstone National Park, withstanding temperatures close to boiling; others thrive half a mile deep under the Arctic ice.

Appearing at a time when the planet looked vastly different, bacteria have developed impressive abilities that enable them to fight and survive. Consider the fact that bacteria initially developed when there was little oxygen on our planet.[2] When some bacteria started releasing oxygen, new bacteria evolved that were far more efficient at using that oxygen to their advantage.[3]

Their pursuit of advantage—whether by preserving the host or by killing it—is constant, inevitable, Darwinian. Faced with an endless competition to survive and reproduce, over time bacteria have developed a highly sophisticated, multilayered defense mechanism that combats external threats and attackers. This defense mechanism works to our advantage when good bacteria produce chemicals that help our immune system fight infection, not just in the gut but also in the lungs and in the brain.[4] Millions of bacteria living in our gut ensure digestion and uptake of nutrients from our food. But "attackers" also include antibiotics (the term comes from two words simply meaning "against microbes"[5]) that we've designed to target and kill the microscopic but mighty life-forms that can just as easily harm the very body in which they reside.

Think of antibiotics as highly specialized weapons that target disease-causing bacteria rather than other cells in your body. Antibiotics occur naturally, and scientists have further enhanced these sophisticated weapons with two goals in mind: to kill the harmful bacteria or to stop it from replicating.[6] Do either, and patients infected with a life-threatening disease from a bad bacteria living inside them have a better chance of surviving.

Now consider the continuously evolving bacterial defense mechanism that threatens the potency of even the best antibiotics today.[7] Bacteria have an outermost defense system, the cell wall, which functions like a heavily fortified castle. Behind it is another wall, called the inner membrane. Like forces arrayed against a castle, antibiotics intent on killing the bacteria can try to poke holes in the defensive cell wall and membrane in a major frontal assault. Some antibiotics stop bacteria from building a cell wall altogether. If it can't destroy the walls, or stop the bacteria from building the walls, the antibiotic opts for going under the radar deep inside the bacteria's interior. The antibiotics use the natural pores and openings of bacteria or diffuse through the lipid membrane to enter. Once inside, the antibiotic has one main goal: to attack the command and control center of the bacteria, a complex but irregularly shaped region called the nucleoid. This nerve center is the bacteria's soft spot. Bacterial replication and information machinery, its DNA, is found in the bacteria's nucleoid zone. The antibiotics have this region in their crosshairs.

Over millions of years, the bacterial system has evolved continuously to defend against antibiotics trying to break through its walls. Bacteria do this through genetic mutations, some of which are random, and some of which they acquire from other foreign bacteria. These mutations are passed down by parent bacteria to their progeny and give them the ability to defend themselves against an antibiotic attack.

The first line of defense, provided by mutations, is formidable. Any antibiotic that is a threat needs to pass through the two barriers—the wall and the membrane. Consider, for example, bacteria that are resistant to the antibiotic vancomycin, one of the "last-line" antibiotics used to treat life-threatening infections such as methicillin-resistant *Staphylococcus aureus* (MRSA), one of the most serious and feared drug-resistant infections in hospitals.[8] The vancomycin-resistant bacteria can make a cell wall that is completely different in structure than the one the drug can recognize. The result? Vancomycin bounces off this new, unrecognizable cell wall and can't do its job.

A bacterial cell can also tighten its borders, reducing the permeability of its walls. As a result, the bacteria can stop or severely restrict the amount of a given antibiotic that gets into the nerve center. And if only a small amount of antibiotic gets in, it is much less likely that it will kill the bacteria or prevent it from replicating.

An antibiotic that effectively breaks through the barriers now confronts a *second* line of defense. Bacteria have one of the most sophisticated mechanisms of cleaning up and expelling threats. The operation uses what scientists refer to as efflux pumps.[9] The pumps work like reverse vacuums. These tiny pumps are located on the cell membrane, and they push out the antibiotics. In some cases, specific mutations in the bacteria's DNA can produce lots of these antibiotic-clearing pumps.

And the pumps are not the last line of defense either. If the antibiotic evades the pumps, the bacteria also have enzymes that act like big cleavers that chop up the antibiotic, rendering it harmless. The antibiotic works only if it is intact. One of the most well-known "cleavers" is β-lactamase.[10] It attacks and chops up beta-lactam antibiotics, which is one of the largest and most widely prescribed families of antibiotics. Penicillin and its derived compounds are in this family, and they are useless if they are chopped to pieces.

The bacterial defense mechanism has another strategy as well: it can make an antibiotic impotent by loading it up with additional cargo. It adds chemical groups to the antibiotic molecule itself, which makes the antibiotic too big to pass through the crevices and crannies it needs to navigate in order to reach its target, the nucleoid zone. Antibiotics need to be a certain size, shape, and form to reach their targets. Think of them as small missiles that need to land deep inside a fortress to access an unguarded area before they explode. If you increase the size of the missiles so they can't reach their targets, you render them useless. That is exactly what some bacteria do.

And there are other bacterial defenses that are even more striking: some antibiotic-resistant bacteria can change the structure or shape of the target. The incoming antibiotic, which is on the lookout for a certain shape and size, can't recognize the target—so it can't complete its task.

Bacteria also enjoy a less visible benefit. They manage all of their evolving of ever more advantageous defenses and resiliency without laboratories, cross-nation collaboration, funding, or the luck of generations of scientists each furthering the advances and insights of the previous generation. Bacteria enjoy far simpler chains of decision making. The function of bacteria—put simply, to take in nutrients and replicate—depends on a chain of command. The bacterial DNA is situated in the nucleoid, the irregularly shaped region inside the bacteria.[11] The DNA has all the information needed for basic processes from replication to metabolism. More fundamentally, this DNA also has the information that creates proteins inside the cell. Proteins, made up of molecules called amino acids, are the workhorses of cell function. Proteins carry out important functions such as transport of nutrients inside the cell and synthesis of important molecules.

Some antibiotics target this chain of command—from DNA to proteins—aiming to disrupt this natural process, which will lead to the bacterial cell's death. To avoid this, some bacteria have created alternative chains of command—that is, they have created alternative proteins to carry out the necessary function needed for survival and replication. The antibiotic ends up targeting the original proteins, not the new ones, leaving the bacteria unscathed. (MRSA is an example of a bacterium that has a new pathway to survive when under assault by the antibiotic methicillin.)

The multilayered bacterial defense mechanism—one of nature's oldest creations, ever evolving, ever surprising—has learned to stay a step ahead of us at every single point in our history together. The consequences for humankind have been catastrophic. At the current rate, when our antibiotics are fast becoming impotent, they are likely to get much worse. Should that happen, routine procedures like C-sections or outpatient surgeries could lead to untreatable infections.[12] It may be the 1918 flu all over again.

CHAPTER 2

FIFTY MILLION DEAD

It was September 1918, and Lieutenant Governor Calvin Coolidge of Massachusetts, who was destined to become president of the United States five years later, signed a dire proclamation. Based on discussions among members of the leadership team that included the governor of Massachusetts, the US surgeon general, the health commissioner, and the division head of the American Red Cross, the proclamation addressed the horrors of the Spanish flu, which was taking the lives of nearly one hundred Bostonians per day.[1] With the state's premier medical staff in Europe to aid American troops fighting in World War I, the document asked for every able-bodied person in the state with any medical training whatsoever to offer his or her services in fighting the epidemic. All schools, parks, theaters, concert halls, movie houses, and lodges were closed indefinitely. Even appeals to God were curtailed: the churches were closed for a period of ten days, or until the situation was under control.

In less than a month, nearly 3,500 Bostonians had been affected. They represented a small fraction of the more than 50 million people worldwide who died of the flu, which lasted about a year and ultimately infected 500 million people. In India, nearly 18 million people died—one observer noted that the holy Ganges was swollen with dead bodies.[2] The ancient city of Mashhad in

Iran lost every fifth person.[3] Across the Pacific in Samoa, the death rate was close to one in four.[4]

While the world remembers the Spanish flu as the killer, most people didn't actually die of the viral disease. They died of complications due to pneumonia, a bacterial infection.[5] The flu virus weakened the immune system, providing an opportunity for the pneumonia bacteria to enter and thrive. In the absence of antibiotics to kill the bacteria, pneumonia proved to be a death sentence.

Fascination with the symptoms of pneumonia goes back at least a millennia, with the Greek physician Hippocrates himself taking an interest in the subject. One of the best descriptions of the symptoms comes from Maimonides, the twelfth-century Sephardic Jewish scholar, renowned philosopher, and perceptive physician, who was born in the Andalusia region of Spain.[6] His talents were all the more remarkable given that he and his family, and the Jews of the Mediterranean, were caught up in the crosscurrents of politics, religion, and competing powers. It wouldn't be the last time advances, scientific or otherwise, would be subject to the whims of other ambitions. Surviving exile and persecution, Maimonides wrote what remains a remarkably accurate account of the disease's assault on the human body: "The basic symptoms which occur in pneumonia and which are never lacking are as follows: acute fever, sticking [pleuritic] pain in the side, short rapid breaths, serrated pulse and cough, mostly [associated] with sputum."[7] Maimonides's treatise on pneumonia continued to be used as a gold standard by medical professionals until the nineteenth century, before the use of modern tools—in particular, the microscope.

In January 1665, a book published by the Royal Society of London became an instant bestseller. The society, started just five years earlier by the Royal Charter granted by King Charles II, had created a new genre with its first major publication: popular science.[8] The book's title was *Micrographia*. Its subtitle was even more enticing: *or Some Physiological Descriptions of Minute Bodies Made by Magnifying Glasses. With Observations and Inquiries Thereupon.* Its author was a thirty-year-old ill-tempered and brilliant polymath named Robert Hooke, and a major selling point was the volume's collection of vivid illustrations of plants and insects. It also highlighted the instruments that Hooke used to see nature in ways never seen before, and the microscope was the most novel of these new tools. (In another first, Hooke coined the term *cell* to describe the basic microscopic structures that he had seen.)

The book was soon available all over Europe and reached the hands of scientists and nature enthusiasts as well as merchants and tradesmen. In 1671, in the thriving fabric market of Delft in the Netherlands, a young merchant named Antonie van Leeuwenhoek became fascinated with Hooke's illustrations. Growing up, Leeuwenhoek had been a curious boy and knew his way around glass blowing and lens crafting. Inspired, he decided to make his own microscope—one that would be much simpler than what Hooke described.

Instead of using two lenses, like Hooke, Leeuwenhoek heated the best Venetian glass to form thin threads and then, reheating the threads, he made small glass spheres that were about one tenth of an inch in diameter. It was a stunning bit of engineering, and though the young man made hundreds of these lenses, he kept his exact technique a secret, one that has remained a mystery to this day. More significantly, the resolution of the images made possible by these little spheres was significantly better than what Hooke had achieved.[9]

Leeuwenhoek did not have the clout of Hooke, who was a fellow of the Royal Society and an alum of Oxford University. Yet Leeuwenhoek kept conducting experiments with anything that he could get his hands on. He examined the thickness of his skin. He studied the tongue of an ox, he looked at the mold growing on bread, and examined the intricate structures on the surface of lice and bees. But his biggest discovery came in 1676.[10]

On Leeuwenhoek's bookshelf was a flask of water infused with pepper that had turned cloudy over the three weeks it had been sitting there. Leeuwenhoek took drops from the flask and put them under one of his microscopes. He then examined each drop individually. What he found was both bizarre and captivating: "I saw a great multitude of living creatures in one drop of water, amounting to no less than 8 or 10 thousand, and they appear to my eye through the microscope as common as a sand does to the naked eye."[11]

He called these organisms animalcules, meaning tiny animals. And in a report sent to the Royal Society—with whose members, including Hooke, he had been corresponding—he both described and sketched them. The Royal Society found his claims to be preposterous. Leeuwenhoek, on the other hand, was seeing these animalcules everywhere—including on his own tongue and teeth. Despite the ridicule from the Royal Society, he remained stubborn about his discovery. Ultimately, a group of church elders and respected men were dispatched by the Royal Society to verify Leeuwenhoek's claims. Using his own microscope, the merchant showed them his animalcules. There was no doubt that Leeuwenhoek was right. His findings were published by the Royal Society in 1677. A new world teaming with organisms had been discovered.[12]

Among the single-celled organisms that Leeuwenhoek was seeing were bacteria. Little did he know that these organisms, which he didn't name in his work, would turn modern science and medicine upside down. A whole generation of scientists, now fascinated by the microscope and how it could help us all understand life, were studying in prestigious labs all over Europe. Botanists and zoologists were intrigued by the life beyond what the naked eye could see. New techniques to see live tissues, and the structures within them, were fast becoming the norm among surgeons and pathologists. And among those early adopters who were using microscopy to study disease was a man named Edwin Klebs.[13]

Restless, highly sensitive, and often combative, Edwin Klebs was an unusual scientist for his time. He was born in the mid-nineteenth century, an era when science was becoming a serious profession, maturing from a hobby or an indulgence. Those who took on this new profession were called scientists, and not just natural philosophers. While most renowned scientists of this period worked in the labs in Western Europe, Klebs, who was born in Prussia, worked not only in Switzerland, Germany, and the Czech Republic, but also in Asheville, North Carolina, and at Rush Medical College in Chicago.

It was in 1875, while working in Prague, that Klebs reported seeing bacteria in the lungs and airways of patients who had died of pneumonia. It was, in fact, a revelatory discovery that the mercurial Klebs didn't make much of at the time. The germ theory—the idea that disease is caused by microbes and germs and not the wind or water—was still nascent and controversial, and Klebs moved on to new labs and different problems. While he didn't push further with his work on bacteria, and his interest moved in new directions, two scientists working on either side of the Atlantic picked up where Klebs's work on pneumonia had left off.

Brigadier General George Sternberg[14] of the United States Army and Louis Pasteur in Paris performed experiments in 1881 that were nearly identical to each other. Sternberg was an army surgeon and an amateur paleontologist. The focus of his career went back and forth between fighting Native Americans and conducting bacteriological research. In 1881, after a bout with yellow fever, he found himself investigating the cause of mosquito-borne illnesses—particularly malaria. It was during the course of this research in New Orleans that he conducted a series of experiments in which he injected rabbits with his own saliva. The animals exhibited pneumonia-like symptoms and died within a couple of days. Sternberg tried the same experiment by injecting water, wine, and saliva from other colleagues. None of these substances produced pneumonia-like symptoms in the rabbits. In the postmortem analysis of the rabbits, Sternberg saw bacteria in the rabbits' blood.[15] The discovery was coincidental, as Sternberg was among the lucky few who carry pneumonia bacteria in their mouths but do not develop the disease.

On the other side of the Atlantic, Pasteur was conducting almost identical experiments, except his source of infection was not his own saliva but that of a child who had recently died of rabies. Pasteur saw the same type of bacteria in the blood of *his* rabbits, bacteria that were long and oval, with a pointed end.[16] Pasteur was faster than Sternberg in publishing his results. He called the bacteria *microbe septicemique du salive*, and Sternberg, cognizant of the fact that Pasteur had already published his findings (and being a stickler for the protocol of the time), called his bacteria *Micrococcus pasteuri*.

Whether the bacteria these men had found was causal to the symptoms of pneumonia, however, was not yet resolved. That fell to a different pair of scientists, both from Germany.

Carl Friedländer was still a young man when he became fascinated by pneumonia. Unlike Pasteur and Sternberg, who put the blood of their subjects under the microscope, Friedländer looked at slices of the lung under the microscope instead. He reported seeing a distinctive bacteria.[17] These bacteria were like a capsule, with a shell around them. These, he declared, were the cause of pneumonia. His claim was bold, and it immediately brought controversy and challenges. Friedländer's own research results were far from conclusive. When Friedländer infected rabbits with his encapsulated bacteria, they failed to develop the disease. Mice, on the other hand, were highly susceptible and would develop the disease immediately. Guinea pigs fell right in the middle, with only six of the eleven that Friedländer infected dying of pneumonia. With such contradictory results, his findings were met with skepticism.[18]

Another physician, Albert Fraenkel, was working in a tuberculosis sanatorium in Germany's Black Forest, and he was also intrigued by the problem. The animal studies he conducted brought him to a conclusion different from Friedländer's. Fraenkel had isolated bacteria from the lung of a thirty-year-old man who had recently died of pneumonia, and when he infected rabbits with it they consistently died in the lab. Guinea pigs, again, gave mixed results.[19]

Another result was more profound. When Fraenkel observed his samples under a microscope, he observed bacteria shaped differently than what Friedländer had reported. So while there was increasingly broad agreement on the fact that bacteria caused pneumonia, there was sharp disagreement about which of the two bacteria, Friedländer's or Fraenkel's, was the real cause of the disease.

As the debate continued, Fraenkel became more aggressive and hostile toward Friedländer and made a number of personal attacks on him. The acrimony continued to the point that Friedländer wrote, in the decorous language of the age, "Of the manifold personal attacks and remonstrates which Fraenkel has directed against me in different places of his work, let them cease. I do not hold them fitting."[20] Fraenkel, however, did not change his tone.

Resolving the discrepancy between the results fell to yet another scientist, who happened to be a graduate student and protégé of Friedländer. His name was Hans Christian Gram. A physician originally from Copenhagen, he spoke German with a Danish accent and had earned his adviser's trust. Friedländer called Gram "my esteemed friend and co-laborer."[21] It would fall to Gram to determine which scientist—Fraenkel or Friedländer—had identified the bacteria responsible for pneumonia.

In the lab, Gram spent his days preparing slides for the microscopes, which required that he use various coloring methods so that some parts of the tissue would become more prominent and easier to see. This work was long and painstaking, and Gram was bothered by the fact that traditional methods of coloring the samples often left multiple parts of the tissue the same undifferentiated color. When the sample was viewed under the microscope, it was difficult to see whether there were any bacteria or not. Gram was determined to improve the process and, consequently, the results. He had read the work of Robert Koch, a leader in the field of bacteriology, and of Paul Ehrlich, a highly prolific scientist and a Koch protégé. Armed with new insights from other German labs at the forefront of dyes and biological staining, Gram set to work.

Gram started off by following the tissue-coloring methods developed by Ehrlich. The initial results were failures. For the research on pneumonia that Gram was undertaking in Friedländer's lab, Ehrlich's methods—which stained the tissue with a particular dye called aniline-gentian violet—did not work. As a result of sheer serendipity and some trial and error, Gram eventually immersed his tissue samples with Ehrlich's stain for three minutes and then for an additional three minutes in a mixture of 1 part iodine, 2 parts potassium iodine, and 300 parts distilled water. This colored the whole sample.

But Gram was not done. He then immersed the sample in absolute alcohol for thirty seconds. What he saw was surprising—nothing except the bacteria retained color. The bacteria were now deep violet and easy to spot under the microscope. What if Gram could color the colorless part of the tissue to highlight the difference even more? That way, there would be no ambiguity. To his surprise, a dye called Bismarck brown worked and colored the whole tissue—except the bacteria.

Friedländer was impressed. He wrote about Gram's method in his book *The Use of the Microscope in Clinical and Pathological Examinations* and had this to say about bacterial cells: "While everything else remains perfectly colorless, they, on the contrary are dyed an intense blue, so that almost every individual in the section must at once strike the eye of the beholder."[22] Through his familiarity with the work of others, along with some luck and perseverance, Gram had come up with a substantial improvement to Ehrlich's method and could now clearly show the bacteria in tissue samples. It also meant he could resolve the debate between Fraenkel and Friedländer.

Gram tested his method on the tissue samples of patients with fatal pneumonia. He revealed that Friedländer and Fraenkel were indeed seeing two different kinds of bacteria in the samples.[23] Fraenkel's bacteria would appear deep blue, almost violet-colored when infused with the stain Gram developed, while the bacteria Friedländer saw did not. Gram's stain showed that the two scientists were seeing two different types of bacteria that were causing distinct types of pneumonia. In a way, both Friedländer and Fraenkel were right, but the more common pneumonia was due to the bacteria that Fraenkel identified. Much later, one of the two bacteria was named *Klebsiella pneumoniae*, in honor of Edwin Klebs, and the other *Streptococcus pneumoniae*. *Klebsiella* is the same pathogen that led to the incurable infection of the woman in Nevada in September 2016.

Toward the end of the nineteenth century, Gram's method became the standard with which to classify all bacteria. If the bacteria took the deep blue color, then they were classified as one type (positive); if not, they were put in the other category (negative). Gram was immortalized for his achievement: bacteria that retain a deep blue color are called Gram-positive, and those that do not are referred to as Gram-negative.

The great questions regarding the appropriate treatment for pneumonia wouldn't be resolved for decades, and millions more would die in the interim. And it was decades later, in the twentieth century, that scientists discovered the real reason why certain bacteria keep the deep blue color and others do not.[24] But with Gram's method, the most fundamental method to classify bacteria had been found, and it remains in use to this day. Diseases such as plague, typhoid, and cholera were caused by Gram-negative bacteria—the same type that Friedländer witnessed. The most common forms of pneumonia, as well as strep throat and anthrax, were all caused by bacteria that retained the deep blue color—the Gram-positive bacteria. Today, any discussion of antibiotics or antibiotic-resistant bacteria starts with a simple question of classification: Is it Gram-positive or Gram-negative? It's an homage to the unassuming man from Copenhagen.

Over the ensuing decades many scientists would engage with the effort to identify, find, isolate, and treat bacterial infection. They would do so with rising sophistication, sometimes in competition with one another and sometimes in collaborative efforts. Sometimes the goal was to heal the wounds of those in battle or those suffering miserably in the TB ward, and sometimes the mission was to find the next blockbuster drug. The scientists would find inspiration in samples from far-off jungles rich with organic matter, in the guts of isolated tribes, and in the dirt samples taken close to the lab. The bacteria were paying close attention to the human endeavor, and our discovery of new generations of antibiotics propelled them to further adapt and evolve.

CHAPTER 3

TIME AND SPACE

Gerry Wright grew up in Northern Ontario, a region renowned for its vast wilderness and sparse population. He was fond of the natural beauty surrounding him. Specifically, Wright was fascinated with soil and the treasures within it. It was a fascination he would follow all the way to becoming a professor of microbiology at McMaster University.

He knew that soil was a rich resource. Some of the biggest antibiotics of the twentieth century, from vancomycin to streptomycin, had come from soil samples. Soil scientists had known for decades, since the early 1930s, that bacteria in the soil are in a perpetual state of war with other members of the bacterial species.[1] Over millennia, bacteria have been making sophisticated antibiotics to kill their competition. How was it, he wondered, that soil bacteria, the resource for so many of our antibiotics, managed to survive? If only a small group of bacteria produce these heavy-duty antibacterial drugs, they should be the only ones surviving. How can any other bacterial species that did not have the ability to make the antibiotics ever survive?

In 2006, a study at Wright's lab in Ontario showed something quite shocking.[2] Wright had been sampling soil from all across Ontario, to study various kinds of bacteria and their naturally existing defense mechanisms. His team found that most of the bacteria in the soil close to his lab were resistant to most of the frontline antibiotics. These bacteria survived because the antibacterial arsenals of other bacteria had no effect on them. The antibiotics just bounced off, as if hitting a high and impenetrable wall. In addition, these resistant bacteria did not cause disease. They were, for all intents and purposes, just minding their own business in the soil. What Wright and his team had apparently shown was that some bacteria that don't produce antibiotics also have a mechanism to protect themselves from attacks by their fellow, antibiotic-producing bacteria. In other words, bacteria that were nonproducers of antibiotics had developed sophisticated resistance mechanisms. This is like a country that has a strong military but little interest in attacking or ability to attack its neighbors.

The paper generated excitement about a new discovery but also met stiff criticism. Many scientists questioned Wright and his team's conclusion that the bacteria had developed resistance on their own. Maybe the bacteria were resistant due to excessive human intervention, perhaps through the dumping of antibiotics into the environment. After all, antibiotics were widely used in agriculture, which was prominent in Ontario. Couldn't the flow of antibiotics through sewage and water channels have contaminated the soil? Wright knew Northern Ontario like the back of his hand. He knew that the area was not contaminated and that the bacteria in his study were resistant for other reasons. There was no reason to expect a high exposure of antibiotics in that region where his soil samples had come from. The problem: he didn't have proof.

And there the matter stood until 2008. Wright was in San Diego attending a conference on microbiology. It was a meeting he remembers well. The Southern California sun sat in a cloudless blue sky, and with its beaches and waterfront, and the expanse of the Pacific Ocean stretching out from its docks and piers, San Diego was absolutely beautiful. But the conference stands out in

Wright's memory for a different reason. A scientist named Hazel Barton presented a paper about the behavior and properties of bacteria found in ancient caves, far from any human activity. The paper was titled "Much Ado About Nothing: Cave Cultivar Collections."[3] This was Gerry Wright's introduction to Barton, a professor at the University of Akron in Ohio. Mesmerized by her presentation, Wright knew instantly that he had found the perfect research partner to answer the big question on his mind: How long have the soil bacteria been resistant?

Barton agreed to work with him to figure out if bacteria could become resistant on their own, even when they have never been exposed to antibiotics. And she knew the perfect place to undertake the research.

In 1984, over the Memorial Day weekend, while other people were busy enjoying their barbecues and preparing for the start of summer, Dave Allured, an engineer in Colorado, was chasing his big dream.[4] He was hoping to discover a new cave. Allured and four of his friends had recently been granted a permit to visit a potential cave site in the Guadalupe Mountains in New Mexico. To reach the site, they started their journey in Colorado and drove south. They drove to the Guadalupe Mountains and hiked in from the nearest access road. They were tired but excited.

Their initial investigation of the site proved extremely promising; they were greeted by gusts of cold air coming from the mouth of a cave where Allured proposed they dig. It took until November to secure the necessary permits from the national park to do so. This time, while their friends and family were enjoying their Thanksgiving turkeys with all the fixings, Allured and his colleagues were digging in the chilly weather. By the end of the holiday weekend, they had made enough progress to know that they would come back and continue their work the following Memorial Day weekend. That expedition went well, but it wasn't until November 1985, with the team now swollen to thirteen, that they had a hunch they were getting close. The whistling gusts of wind coming from where they had dug meant they were very close to a deep cave.

The team again commenced digging in the spring of 1986. Over the last expeditions, the dig had reached about thirty feet long, and Allured and two colleagues, Neil Backstrom and Rick Bridges, were determined to push farther. The task had grown more precarious. Boulders were blocking the passage and the risk of a collapse was increasing, but the trio persisted. As they dug, they found cave pearls, giant flowstone structures that looked like bells, cave chandeliers, and an underground crystalline lake. It was all breathtaking. Slowly, they progressed, and the structures gave way to a big pit. They could not see the bottom.

The team estimated that the pit was 150 feet deep. It was, in fact, much larger and deeper than they could ever imagine. Since May 26, 1986, when Dave Allured and his team made their discovery, explorers have mapped more than 136 miles of cave passages, making the Lechuguilla Caves one of the largest cave systems in the world. It remains the deepest limestone cave in the world. And it proved to be a bacterial treasure trove.

It was here that Barton conducted the research that sparked Wright's interest at the conference. Combining her two passions—microbiology and caving[5]—Barton worked with Wright for over three years on an ambitious and at times life-threatening project that aimed to answer the question: Can there be antibiotic-resistant bacteria living deep inside a cave that has never seen any human activity? It required Barton to go far into the depths of the Lechuguilla Caves collecting samples of biofilms, which are multicellular communities of bacteria that are often attached to surfaces. In the course of the research, she would go into a region of the cave that's called Deep Secrets.[6]

Deep Secrets is about 1,300 feet from the surface of the Earth. To reach it requires traversing terrain that is treacherous and terrifyingly beautiful. The meticulous Barton, wearing her caving gear and carrying minimal supplies in her backpack, collected ninety-three different strains of bacteria from this nearly inaccessible cave. She brought them back to the surface, and, using the latest tools and technologies available, Barton and her colleagues carefully screened them against most known antibiotics to find out if these bacteria, which had never seen any commercial antibiotic or any human activity, were resistant to the antibiotics that were commercially available and used in hospitals. The bacteria from the Lechuguilla Caves, Wright and Barton discovered, held their own deep secret.

Though the bacteria that Barton collected was from a part of the Earth that had been isolated from human activity, it was resistant to some of the most potent antibiotics, including daptomycin, which is used for treating MRSA.[7] To the skeptics who doubted Wright's original findings, he and Barton now had compelling evidence. Bacteria from a location that had been cut off from all human civilization for nearly 4 million years was antibiotic resistant.

The research showed something even more peculiar: the resistance genes carried by the bacteria and the mechanisms by which they defended against the antibiotics were similar to what we see in patients who are resistant to chloramphenicol,[8] a drug still used in many parts of the world to treat typhoid. This news was turning the whole field of microbiology on its head. The assumption, up until that point in 2012, had been that resistance was largely driven by human activity—excess, greed, ignorance, and hubris. The general understanding was that bacteria develop specific genetic mutations in the presence of antibiotics. These mutations allow bacteria to create new defensive mechanisms against antibiotic assault. Some of these mutations are natural, either random or in response to the antibiotic attack from other bacteria. But many are also a result of excessive exposure to antibiotics discovered by scientists and produced by pharmaceutical companies. How could bacteria that had never seen any human activity, and predated humans by millions of years, have the same resistance patterns that doctors witnessed in hospitals?

Barton and Wright dug in even more. They wanted to be absolutely sure of their results. They chose to compare a cave bacterium that had been isolated from any and all human activity with a cousin bacteria on the surface of the Earth with all of its exposures to human and animal activity.

The team chose a particular bacteria from the cave called *Paenibacillus* sp. LC231 and compared it to surface bacteria from the same family, *Paenibacillus lautus* ATCC 43898. To their surprise, LC231 was resistant to most clinically useful antibiotics. Out of the forty antibiotics they tested, the bacteria was resistant to twenty-six.[9] But there was more. When the team investigated the mechanisms of resistance, they found that several of the mechanisms by which LC231 was showing resistance were familiar and well documented in other bacteria, but there were at least three new mechanisms of resistance that had never been reported before.

The team had shown, definitively, that the bacterial-resistance mechanism was very old and predated human activity or the marvels of modern medicine. It had little to do with the overuse or abuse of antibiotics by humans. The bacteria in the cave had developed their own defense against an antibiotic attack, and these bacteria had been remarkable at conserving this defense machinery for millions of years.

The story was big—and a potential boon to those who had been arguing that human activity does not affect nature. Some in the industries and drug manufacturers grabbed hold of these findings to argue that if bacteria have been producing their own resistance mechanisms, and had been doing this well before any antibiotics were made, there was no need to stop using antibiotics in, say, agriculture and food production. What difference did it make if, globally, three times more antibiotics were used in livestock than were given out to people in clinics and hospitals? Clearly the bacteria were doing what they had been doing for millions of years, independent of human behavior. Resistance was real, they argued, but had nothing to do with industrial agriculture.

Wright believed this response was shortsighted.[10] What his years of research revealed, starting with the soil taken from Ontario and continuing into the reaches of the Deep Secrets, was that bacteria would develop antibiotics on their own, and then other bacteria competing for the same resources would develop resistance to protect themselves from the bacteria that had the antibiotics at their disposal. It was a balanced competition between antibiotic-producing bacteria and antibiotic-resistant bacteria that had played out over millennia—except now, with overuse of antibiotics around the world, this balance had been disrupted in the soil, in water streams, on doorknobs and bedsides, and of course in hospitals. Drug-resistant bacteria now had a clear edge and were thriving without any real competition. Wright was convinced that human activity was responsible for this disruption of equilibrium. But his results also created a glimmer of hope. The bacteria were active in producing not just new defense mechanisms but also new molecules that could disarm resistant bacteria. There were plenty of undiscovered antibiotic molecules hidden deep inside the treasures of the earth, waiting to be discovered.

CHAPTER 4

FRIENDS IN FAR PLACES

For Gerry Wright, it was soil, but for Gautam Dantas, it was the Yanomami tribe of the Amazon.[1] More precisely, it was the microbiome of the Yanomami tribe and, initially, the origin story of the *Helicobacter*, a genus of Gram-negative bacteria that thrives in most human stomachs.

An arbitrary border between Brazil and Venezuela goes through the thick jungles of the Amazon. Here the vegetation becomes so dense that sunlight is a precious resource. Every inch of ground is contested real estate. Plants tend to grow tall, and they have thorns sharp as daggers to keep competitors away. Serpents, seagulls, armadillos, wild pigs, and jaguars have roamed this jungle for millennia and have shared it with their evolutionary cousins: humans.

Anthropologists believe that the Yanomami tribe has lived here for about eleven thousand years. They are known for their communal living—their large, circular buildings and centralized community area is often reserved for rituals and festivals. And this tribe has also been sought out for scientific research.

The tribes in the southern part of Venezuela and northwestern Brazil, including the Yanomami, remained largely isolated from the rest of the world until the 1970s, when new development and "integration" policies of the Brazilian government resulted in catastrophic massacres, displacement, and suffering. By the

1990s, mining companies saw the region as their own personal El Dorado. Fueled by a combination of government apathy, kick-backs, and lobbying, the companies pushed deep into the jungle, bringing the inevitable new diseases with them. The Yanomami suffered again, along with the Amazon, which slowly gave way to the onslaught of industrial development.[2] But some of the tribes, particularly those on the Venezuelan side, remained beyond the reach of change and destruction. Most of them remained un-known, until a team of army personnel in 2008, while aboard a helicopter, saw a group of Yanomami people in an uncharted village. Scientists would soon find out the secrets that their gut bacteria held. But first, the Yanomami people, shielded from civ-ilization and never exposed to modern medicine, had to be pro-tected against potential disease.

The bustling metropolis of Mumbai is about as far from an Am-azonian rain forest as one can imagine. For Dantas, the "jungle" in densely populated Mumbai was constructed out of concrete, not trees, and the background noise was that of traffic-choked streets, not exotic birds. In his tenth-grade biology class, fas-cinated by a talk given by an expat American teacher, Dantas made a split-second decision. Inspired by tales of the biochemical potential of marine organisms, he decided that he would one day obtain a PhD in biochemistry. In pursuing this goal, a series of serendipitous events launched him into an educational odyssey.

His first stop was an international boarding school in Kodai-kanal, a small town in South India. From there, he was off to Ma-calester College in St. Paul, Minnesota, his stepping-stone to the University of Washington in Seattle, where he finally got that PhD. And armed with the degree and all the knowledge he had accu-mulated over the decades, Dantas next went to Harvard Medical School. His zigzagging around the country carried through to his academic disciplines as well: he studied organic chemistry and protein engineering and biofuels.

Biofuels were a hot topic in the early 2000s, and one epicenter of interest was at Professor George Church's lab at Harvard. Church was known for his highly unusual and creative approaches to science, and at that moment the people in his lab were busy looking at the ways in which bugs could be used as biofuel.

One day, during his postdoc at Harvard, Dantas and his fellow lab members set up a simple experiment that would unexpectedly transform Dantas's research interests. Like generations of scientists before them, they turned to soil for its infinite reservoir of microbes. They collected soil samples from a number of different locations around the United States and isolated the bacteria within them in hopes of finding material from which they could create useful biofuels. The first decade of the 2000s was full of uncertainty about global oil and gas markets, and biologists, chemists, and chemical engineers were actively trying to find alternative fuel sources, including biofuel from organic matter and microbes such as bacteria. The traditional barriers between science were also starting to disappear, with new interdisciplinary methods being adopted and implemented across academic institutions. To culture the bacteria, Dantas and his colleagues tested various plant-based compounds for their potential as bacterial food sources, which in turn could be converted into biofuels. In their control experiments they included antibiotics at concentrations that were far beyond toxic. In fact, the levels of antibiotics were many times higher than the maximum doses we give to people with major bacterial infections. Dantas was certain that any bacteria in these control experiments would be eradicated. But something different occurred.

Dantas and his peers were shocked to find that the antibiotics didn't kill the bacteria: instead, some of the bacteria actually gobbled them up. It was bizarre. These bacteria were not only surviving—but thriving—on these antibiotics. Dantas couldn't believe what he was seeing. How did these bacteria engage with their environment, and how did that environment influence their behavior? As Dantas was sorting this out, the scientific community's interest in the human microbiome was skyrocketing. It encouraged Dantas to ask new questions: Could the overall relationship between bacteria and the environment be playing out in the human gut as well?

Dantas was hooked on bacterial behavior and, more important, on microbial genomics. By 2009, as Dantas started his own lab at Washington University in St. Louis, he focused his interests on the microbiome. He attended a conference where he met Rob Knight, then a professor at the University of Colorado in Boulder, who was known for his sophisticated computational studies of microbes and their genomes. The two scientists had much in common, but crucial to their relationship was an interest in developing tools to explore the behavior of microbes in various environments.

Knight and Dantas started talking about an experiment that they believed would shed light on the microbial world, but they had very specific parameters in mind. They wanted to see if they could look at the microbiome of people who had not had any exposure to antibiotics. What would that look like? Could that experiment even be done? Dantas doubted it, perhaps reflecting on the concrete jungle of Mumbai. But as luck would have it, Knight had been collaborating with the scientist Maria Gloria Dominguez-Bello.

Dominguez-Bello was a microbiologist at the University of Puerto Rico (she moved to New York University in 2012 and became a professor at Rutgers University). She had a long-standing interest in the microbiome. For most of her career, she had been busy studying *Helicobacter*. *Helicobacter* are a curious bunch. We all carry them in our gut—well, at least two-thirds of the world population does. For most people they are benign and harmless forms of bacteria—and scientists believe that the presence of these bacteria in the stomachs of children can reduce allergies as well as asthma. Occasionally, however, *Helicobacter* can also cause ulcers in children and adults by damaging the inner lining of the stomach and creating a sore.[3] Dominguez-Bello had been studying *Helicobacter* for some time, investigating a question that was often asked of her in her native Venezuela. Did the Europeans bring *Helicobacter* to the New World? Or was it already there when they arrived?

It was well known that the Spanish colonizers had brought with them a whole host of diseases—ranging from measles to smallpox—that had decimated the native populations. But Dominguez-Bello wanted to know if the Europeans were carriers of *Helicobacter* and had introduced this bacteria to the region and, consequently, to the guts of native peoples. Dominguez-Bello began investigating. She knew that the native peoples of South America—also known as Amerindians—had originally come from Asia. She also knew that *Helicobacter* strains from Europe were different from the ones in Asia. This lineage could be shown by lab experiments. If she could somehow study the *Helicobacter* from the gut of native tribes, ones that had not had any exposure to European colonizers, she would have her answer. Working with colleagues in Venezuela and other Latin American countries, Dominguez-Bello showed that the tribe's *Helicobacter* strains were Asian. The European settlers had brought lots of pathogens and diseases to Latin America, but *Helicobacter* was not one of them.[4]

Through her contacts in the Venezuelan government and the Amazonic Center for Tropical Diseases,[5] with whom she had worked since the 1990s, Dominguez-Bello learned that the Venezuelan army had recently sighted a group of Yanomamis from a helicopter during a surveillance mission.[6] The Yanomamis were far from the Amazon River and in a region not previously mapped. It meant that they were a potential treasure trove of scientific data—and that they were highly vulnerable. Understandably, their location was protected by the local government. Even Dominguez-Bello, to this day, does not know the exact location of the tribes. It is kept a secret. And yet there is a persistent danger of the tribe being wiped out if—and, by most reasoned accounts, not if but *when*—it ever were to come into contact with communities that have immunity or have been vaccinated. Because the

Yanomami were never vaccinated, they probably did not have immunity to deadly infectious diseases like measles.

In 2009, a medical mission from the Venezuelan government landed at the site to vaccinate the community from communicable diseases. Following strict protocols to ensure they did not introduce disease, the team found that the small village was composed of hunter-gatherers, and that the tribe subsisted on a diet of wild bananas, seasonal fruits, small fish, and frogs found in the marshes. The nearest medical outpost was two weeks' walking distance. Before the members of the community were vaccinated, the medical team collected fecal, oral, and volar forearm skin samples. And it was these samples that Dominguez-Bello was interested in studying to understand the tribe's microbiome. To gain access to the samples, Dominguez-Bello waded through an extensive pile of paperwork, and it took almost a year to get all the approvals.

Rob Knight connected Dominguez-Bello and Dantas. The samples from Venezuela, which Dominguez-Bello had access to, could provide more information than just the origin story of *Helicobacter*. They could tell whether the native tribes—living far in the jungle and with no history of antibiotic use—were, in fact, naturally carrying any genes of antibiotic resistance.

Dantas's lab got to work. Access to the samples was well guarded (just as Dominguez-Bello experienced), and properly so. They were incredibly rare and incredibly precious. Dominguez-Bello provided the contacts to the proper Venezuelan authorities, but the lab had to do the paperwork. And it was all to pursue a speculative outcome. The bureaucratic hurdles took months to clear, and as they waited, Dantas and Knight made sure that their lab was fully prepared for the eventual arrival of the samples, which landed in St. Louis in early 2013.

The team went into overdrive—and the initial results were both fascinating and troubling. The studies showed that the members of the Yanomami carry bacteria that have resistance to some of the common antibiotics that are currently used in modern hospitals across the globe. But there was something more. The microbiome of the Yanomami had resistance to advanced, sophisticated antibiotics.[7] The data showed that the tribe had resistance to naturally occurring antibiotics, but also to synthetic antibiotics—which, by definition, were not supposed to be found in nature. How had a tribe closed off in the jungles of the Amazon developed resistance to drugs that had been created in pharmaceutical company labs during the 1980s?

Dantas, Knight, and Dominguez-Bello showed something no one had expected—as was the case with Gerry Wright and Hazel Barton. Not only was the bacteria in the microbiome of previously uncontacted people antibiotic resistant, but the Yanomami were also resistant to drugs that were supposedly created in the lab.[8] How could this be? Dantas can only speculate—and two options seemed likely: First, the molecular fragments discovered by Dantas and his team may perform natural functions in the Yanomami bacteria. That they also defend against "modern" advanced antibiotics designed in distant laboratories is simply serendipitous.[9]

However, there is also a second possibility, and it was cause for Dantas's optimism and that of many of his colleagues. Maybe these so-called synthetic antibiotics are not synthetic at all. Maybe what scientists started making in the lab are compounds that nature has been making all along. Perhaps there is a chance that these synthetic molecules are being made naturally by some environmental bacteria in the Amazon. And these compounds, which are a part of the immediate environment, are reaching the tribes and selecting for resistance in the microbiome of the local people. This may mean there is a vast untapped reservoir of new drugs and chemicals present in various microbiomes around the world.

Scientists and microbiologists have believed for decades that the soil may hold an infinite reservoir of bacteria that are able to produce potent antibiotics. The early successes in the 1940s, and the honeymoon period for drug companies in the 1950s that were discovering new antibiotics from soil samples, attest to this fact. But as the pipeline for new drugs started to dry up in the 1960s, scientists started to look elsewhere.

Dantas, Dominguez-Bello, Wright, and others have shown that the hunch of early soil scientists was right. There is, indeed, reason to be optimistic. The discovery of the resistant cave bacteria and the Yanomami genes suggest that we have not yet exhausted nature's reserves. Bacteria that can become antibiotic resistant, in the absence of modern medicine, suggest that there are antibiotics out there, not yet known to us, developed and used by bacteria in their own battle for Darwinian survival. And looking for these potent chemicals is going to require ever more sophisticated techniques.

CHAPTER 5

NEAR THE SEED VAULT

Gerry Wright was excited about his big discovery. But something was bothering him. He wondered about the arms race between bacteria. Some bacteria are producing antibiotics, and then, to prevent suicide from these deadly molecules, they develop resistance. Some bacteria, in order to survive and prevent attacks from other bacteria, also develop this defense. So how long does this arms race go on? If defense gets to be too strong, what is the advantage of having any offense whatsoever?

Something did not make sense to Wright. If every bacteria was eventually able to resist antibiotics, why were some bacteria still producing them? As more and more studies showed the increase in resistance among bacteria, Wright, a student of nature and captivated by the natural order of things, wondered if there was another angle—a means for some bacteria to weaken the defense of other bacteria, making them vulnerable once again to antibiotics.[1]

To test his theory, Wright went back to the most potent of his resources—soil. He asked his students and colleagues to collect any samples they thought were promising. He was looking for a missing piece of the puzzle that showed that there was still a reason for bacteria to produce antibiotics. The answer came soon. One sample came from a national park in Nova Scotia where one of his students had been hiking during a vacation. The soil sample was analyzed like hundreds of others, but this one ended up being special. In it, Wright and his team found molecules produced by *Aspergillus* fungus, a form of fungus that typically grows on decaying leaves.[2] What Wright's lab discovered was that *Aspergillus* produced molecules that can disarm bacteria's resistance mechanisms. This was a new twist in the arms race. Some bacteria would produce antibiotics; other bacteria, to save themselves, would develop resistance. But then there were molecules, like those produced by *Aspergillus*, that could disarm the resistance, poking holes in the bacterial defenses, making the original antibiotics potent again. Wright had found one such molecule called aspergillomarasmine A, or AMA, in the Nova Scotia soil sample.[3]

So his research took on a new focus. Now, his target was to convert this into a weapon against bacteria that contain a peculiar enzyme that makes bacteria resistant to large numbers of highly potent drugs. The name of the enzyme is New Delhi metallo-beta-lactamase 1, or NDM-1.

Little connects the teeming population and activity of New Delhi, India, and the Svalbard archipelago. The latter is situated high up in the Arctic Ocean, about midway between the North Pole and mainland Norway. Not much grows here. In the seventeenth and eighteenth centuries, the archipelago was a whaling base for Norwegian explorers. Today, Svalbard is known for something altogether different. It is home to a vault intended to preserve human civilization in case of Armageddon.

The Global Seed Vault is located on the island of Spitsbergen on the western end of Svalbard. The vault is an ultra-secure facility nearly four hundred feet under a sandstone mountain[4] and is managed jointly by the government of Norway, an international trust of nongovernmental agencies for food security, and a Nordic consortium focused on genetic resource preservation. The vault itself contains samples of nearly 4.5 million seeds, preserved and ready to be used in case of a global apocalypse. Ironically, the vault may not survive an apocalypse unforeseen by its Norwegian designers: climate change. Due to melting glaciers, the vault is at risk of flooding. So though for practical reasons the site was chosen due to its being far from human civilization, it has proved not out of that civilization's reach. And not just by a changing climate.

In the northwest corner of the island of Spitsbergen (where the seed vault is located) is a glacial fjord named Kongsfjorden. Here, in January 2019, an international team of researchers from the UK, the United States, and China reported the presence of antibiotic-resistance genes in soil collected from multiple locations in Kongsfjorden, including a tiny lake that they called SL3.[5] Among the drug-resistance genes that they found in soil samples from SL3 were genes resistant to a class of drugs that is among the last lines of defense against pathogenic bacteria. This class of drugs is called carbapenems. The research team found the soil samples contained the same gene that Gerry Wright was working on. It was also the *same* one that was found in the carbapenem-resistant infection of the woman in Nevada: the dreaded NDM-1.

CHAPTER 6

KEYS TO KARACHI

Arms races and teetering balances of power aren't exclusive to the world of bacteria. Indeed, how these tensions play out among humans has directly influenced the contest among bacteria since well before the time of Maimonides. And two densely populated, nuclear-armed nations found in the heart of Asia—India and Pakistan—remain perpetually on the brink of war, and possible global catastrophe.

Since the partition of British-governed India in 1947, the India-Pakistan rivalry has shaped generations on both sides of the border. The two countries have survived two major wars, in 1965 and 1971, and smaller local skirmishes in the highest battleground in the world, the frigid glacier region of Siachen. Incursions, acts of terrorism, and a war of words all reflect an enmity running between the countries that is profound and enduring. It was within the backdrop of this rivalry that a British bacteriologist by the name of Timothy Walsh became a celebrity in Pakistan.

Tim Walsh was born in Bristol but moved to Tasmania, the southernmost state in Australia, at the age of thirteen.[1] His love of science was shaped by his father, who was a biologist. Initially thinking that he wanted go into medicine, Walsh changed his mind and ended up with a master's degree in microbiology with a focus on beta-lactams—the class of antibiotics to which penicillin belongs.[2] After completing his undergraduate degree in

Australia, Walsh returned to Bristol and completed his PhD. He moved around in the UK, seeking the right lab: one that would both provide him the necessary professional training and give him the intellectual freedom he craved. After a series of fellowships, he finally landed a job right back at Bristol as a lecturer. He had come full circle but remained restless. While the field of antibiotic resistance was enamored with Gram-positive bacteria at the time, Walsh was stubbornly pursuing his interests in the world of Gram-negative bacteria. In 2006, an opportunity arose in Cardiff, Wales, where the university was interested in recruiting Walsh to a chaired professorship. Cardiff was an hour west of Bristol across the channel. Walsh liked the quiet environment of the university, and the professorship promised fewer teaching and administrative burdens.[3] And so he accepted the job and settled into a routine of research on Gram-negative bacteria.

Early in 2008, Walsh got a call from a colleague at the Karolinska Institute, Stockholm, hoping Walsh could help him. The call concerned an Indian man living in Sweden. The man had recently been to India and had contracted a urinary tract infection (UTI) that continued while he was in Sweden. UTIs are not unusual. Routine lab tests were ordered to isolate the cause. And then: lab tests came back showing that his infection was due to the same kind of pneumonia that had excited Friedländer over 120 years before.

There was more. The bacteria causing the infection had a new gene (producing a genetic code for a novel enzyme), which contributed to the bacteria's possessing high-level resistance to a whole host of drugs. Walsh named the enzyme New Delhi metallo-beta-lactamase-1. The name reflected international convention, which followed the standard combination of the origin of the patient's illness, New Delhi, and the mechanism of resistance, metallo-beta-lactamase.

Despite its name, and the fact that it could confer such widespread resistance, no one paid much attention to this discovery, certainly not in India. And ultimately it mattered little to Walsh what the name of the gene or the enzyme was, or where it was first isolated. The real reason it troubled Walsh was that NDM-1 could confer resistance to some of the most potent antibiotics out there. It could go from one bacterial species to another. It could make *Klebsiella pneumonia*, *Escherichia coli* (typically called *E. coli*), and other Gram-negative infections (like urinary tract infections), completely resistant to penicillin, cephalosporins, and carbapenems. A significantly larger population could be at a very high risk of treatment failure. It was time to take the research to the field.

Walsh and his team traveled to India in 2009 and started collecting samples from patients, sewage, and local water supplies. They worked closely with Indian colleagues who were equally interested in learning the scope of the problem and the epidemiology of this new resistance mechanism. What Walsh and his team discovered left all of them very concerned.

Studies conducted across India revealed that dozens of patients had NDM-1 present either in *E. coli* or in *Klebsiella pneumoniae*. Every sample was checked and tested at UK labs to ensure the reliability of the results. Even as Walsh and his teams conducted their research, reports started to come in that patients in the United Kingdom and the United States were *also* carrying infections with NDM-1. Even more troubling was the fact that the CDC reported that all of the American patients who had NDM-1 had come back from India or Pakistan. Worse, all the US patients who had NDM-1 had one more thing in common. They had been treated for various illnesses at local hospitals in India or Pakistan.[4] That wasn't too surprising—this was at a time when India's medical tourism industry was booming. But the implication was alarming. It meant that there was a resistance mechanism lurking in Indian hospitals—one that conferred resistance to medicine's most promising antibiotics—and it was spreading around the globe.

With the publication of Walsh's discovery in the prestigious journal *Lancet Infectious Diseases* in August 2010, news articles about Walsh's alarming discovery started appearing around the world. A Google search of NDM-1 returned well over 4.7 million hits in two days after the publication of the article. Headlines fueled the global fear of a major drug-resistant outbreak that would be impossible to control: "Scientists Find New Superbug Spreading from India"[5]; "Are You Ready for a World Without Antibiotics?"[6] India's tourism industry took a big hit as people began to wonder if it was even safe to visit the country. This caught the Indian government by surprise, but they reacted almost immediately. The government damned Walsh's study, called it biased and even unscientific. It was deemed malicious propaganda against the tourism industry, concocted by the English, who remained jealous of the success of its former colony. A member of Parliament, S. S. Ahluwalia, captured this sense of injury and denunciation: "When India is emerging as a medical tourism destination, this type of news is unfortunate and may be a sinister design of multinational companies."[7]

Safely back in Cardiff and well established in his career, Walsh wasn't too concerned about the backlash. But the first author of the paper, Karthikeyan Kumaraswamy, was accused by members of the Indian Parliament of colluding with the West and being guilty of a conflict of interest.[8] While pursuing a scientific mystery, Kumaraswamy had become the center of attention in a political storm that was putting the $2.4-billion-a-year Indian medical tourism industry at risk.

The bone of contention, from the Indian point of view, was the naming of the enzyme. NDM-1 was named according to international convention, but connecting the new superbug to New Delhi was taken as an offense by the Indian government, which feared that linking this gene to their country would permanently affect their medical tourism industry.[9]

The government started conducting its own studies, which

reported results that contradicted those from the independent researchers. Critics questioned the validity of the new findings, worrying that they were biased as a result of the governmental interference. Tensions increased within the government, the health industry, and public health organizations. Meanwhile, scientists pointed out that the prevalence of the gene that would encode the enzyme was what really mattered. Its ability to resist antibiotic treatments remained while people argued about its name.

Walsh and his team persisted. Their new studies showed high levels of antibacterial resistance in India's water, sewage, and waste. Analyzing the water and sewage samples from New Delhi, they estimated that nearly half a million people—at least 10 percent of all residents of the city—were carrying NDM-1–producing bacteria in their gut.[10] This would mean that if they fell ill, and their sickness was due to a Gram-negative bacteria like *E. coli* or *Klebsiella pneumoniae* that carried NDM-1, the standard treatment wouldn't work.

When word got out about this discovery, another media storm hit, which included a live televised conference featuring the director general of the Indian Council of Medical Research, Dr. Vishwa Mohan Katoch, who said the latest findings lacked any clinical or epidemiological evidence. The government assured residents that the water was safe for consumption and that the citizens need not worry,[11] even as it quietly started chlorinating their water more aggressively and distributing chlorine tablets. To decrease the likelihood of future such embarrassments, the government quickly changed the rules on sample collecting.[12]

Walsh was becoming persona non grata in India, but Pakistan hailed him as a hero. The reasons had little to do with the fact that he was an esteemed, conscientious scientist and collaborator. No, he was lauded for showing that the water in Delhi was dangerous. He had brought infamy to India. The fact that NDM-1 was not just limited to Indian patients but was of global concern, or that patients in or from Pakistan were also carrying the antibiotic-resistant bacteria mattered little. The Indian medical tourism industry was in trouble, and that was all Pakistan needed to celebrate.

In a formal ceremony held in Karachi, Pakistan, in 2012, Walsh was honored by the mayor of the city. The fact that bacteria ignores borders and nationalist histories simply went unmentioned.

CHAPTER 7

WAR AND PEACE

An unassuming memorial to the worldwide search for a medical miracle, the cure to bacterial infections, currently sits about six miles from central Tokyo. There you will find a small shrine, across from the tall buildings that make up Kitasato University Hospital. Like the university and hospital, the shrine also recalls Shibasaburō Kitasato, a world-renowned physician and bacteriologist. Yet on the shrine, Kitasato shares billing: it is called the Kitasato Koch Shrine, named after two scientists who couldn't be more different.

Built in 1920, the shrine has moved twice over the last one hundred years. No matter its location, the shrine is a destination, and a ceremony is conducted at the shrine on May 27 every year. The memorial ceremonies follow the Shinto customs for the dead, and for the first several years after its dedication, the person in charge of leading the religious ceremonies was Kitasato himself.

When I went to visit the shrine in the summer of 2019 I heard that there was a centerpiece to the memorial that contained a single lock of hair—but the hair did not belong to an ancestor of Kitasato. The offering marked the achievements of a man heralded for beating back death. It belonged to a German microbiologist by the name of Robert Koch.

While Kitasato commanded fame in Japan, where he was

rightly lauded as a pioneer in the field of infectious diseases, he traced the origins of his accomplishments to 1886, the year he started working in Koch's lab. In Koch's eyes, Kitasato had evolved from a Japanese who spoke good German to someone who had become indispensable for the lab and the newly created Institute for Hygiene in Berlin.

Echoing that assessment, when Kitasato died in 1931, his students renamed the shrine from Koch shrine to Kitasato-Koch shrine.[1] The shrine moved from the Institute of Infectious Diseases to the newly formed Kitasato Institute. It would change location one more time, after being badly damaged during World War II, an ironic acknowledgment of that war's role in being among the greatest drivers of both bacterial infection and antibacterial innovation and advance.

About three miles from Alexanderplatz in the center of Berlin, on the banks of the Berlin-Spandau Ship Canal, stands the Robert Koch Institute. The stately redbrick building is surrounded by beautiful magnolia trees with pink flowers that blossom in the spring. The nearby canal connects the Havel River to the Spree River, on the banks of which Berlin was first founded. Opened in 1900, the building and the institute it houses, like the shrine in Japan, reflect Robert Koch's stature and global reach.

Koch was arguably the most influential bacteriologist at the turn of the twentieth century. His foray into public health came through a rigorous, uniquely German model of education pioneered in the late eighteenth century. Born into a large mining family in 1843, Robert was the third of thirteen children. He was educated in Göttingen, which in the mid-1850s was the foremost location to study the natural sciences. But for reasons still debated, Koch switched to medicine and became a doctor, graduating in 1866 with the highest honors.

France and Prussia went to war a few years after Koch graduated. It would leave an indelible impression on the young man, who witnessed the horrors of the war as a surgeon working at the front lines.[2] Well after the war ended, the animosity between the two countries continued to shape Koch's worldview. Though the conflict ended as a decided Prussian victory, it left permanent scars across Europe, visible and not, and laid the groundwork for a far more encompassing war.

In the war's immediate aftermath, Koch returned from the front lines to become a country doctor in Wollstein, in present-day Poland.[3] In the late 1860s, Wollstein was a rural area where the vast majority of the farmers were preoccupied with a disease that today sends chills down our spines as it invokes images of bioterrorism and secret laboratories: anthrax. In the late 1800s, anthrax was not associated with warfare or terrorism but was instead a major challenge for farmers concerned about their animals' health and, occasionally, their own health. Caused by a bacteria called *Bacillus anthracis*, there are four forms of anthrax. The deadliest form takes a route through the respiratory system, and the spores of the disease rapidly multiply inside the body. The initial flulike symptoms lead to severe and rapid tissue damage in both the lungs and the lymph nodes. In the absence of treatment, death is nearly certain for this type of anthrax.

Koch knew that animals affected by anthrax could die quickly—and in a visibly painful way. He was determined to find the cause of the deadly infection; he wanted to learn the secrets of anthrax and find a cure for it. Luckily, in addition to being a doctor, Koch was an ingenious craftsman. He was patient, careful, and deliberate. His lab skills were superb, a remarkable gift that set him apart from other doctors. Using primitive slivers of wood, he injected blood from the spleens of anthrax-infected cattle into the tails of healthy mice in order to figure out if the disease seen in different animals had a common cause.[4]

Exact in his practices and measurements, and despite the limitations of his rural environment, he was able to culture and then identify the pathogen for anthrax. The result was revolutionary. Koch had shown that there was a single pathogen that was responsible for every single case of the disease, no matter what form.

Up until that point Koch was virtually unknown in the rapidly growing field of infectious diseases. But Koch knew the bacteriologist Ferdinand Cohn.[5] Cohn had started his training at the University of Breslau, but because he was Jewish, widespread anti-Semitism in Germany prevented him from completing his doctorate degree there. He moved to the more cosmopolitan Berlin and got his doctorate at the age of nineteen.

In a cruelly ironic twist, though he could not be admitted as a student for his doctorate at Breslau due to his Jewish heritage, the institution had no such restrictions on Cohn becoming a professor there. He joined the university in 1849 and remained there for the rest of his professional life, during which time he developed a systematic method of classifying bacteria, one that became a crucial precursor to the method we use today. By the late 1860s, Cohn was considered a reigning authority on microbes.[6]

On April 22, 1876, Koch wrote a letter to Cohn and announced that he had discovered the complete life cycle of anthrax.[7] Cohn was skeptical but intrigued, so he invited Koch to Breslau to discuss the matter. Over a period of three days, Koch demonstrated to Cohn and his colleagues how simple instruments could provide a clear, irrefutable basis for his claim—that anthrax was indeed caused by bacteria, one that could grow inside healthy animals. In addition, if conditions were unfavorable, this bacteria could also form spores and remain dormant until it could once again find a favorable environment in which to grow.

Cohn was convinced, and Koch published his findings in 1876.[8] His instant fame was not due to his discovery of the anthrax life cycle alone—he had also argued for something even bigger: germ theory. This was the idea that diseases are caused by a unique germ, a pathogen. It wasn't "bad air" that caused the maladies of the world, but unique germs that caused disease. A paradigm shift was taking place—diseases that were traced to bacteria needed scientifically sound treatments, not superstitious solutions and folk remedies.

Koch formulated four postulates—now known as Koch's postulates—to argue for his germ theory.[9] In the first two postulates he argued that the disease-causing pathogen must be present in all cases of the disease, and that the pathogen can be isolated from the diseased host. In other words, if someone were to catch a disease by a bacterial pathogen, that disease-causing bacteria must be physically present in the patient, and a doctor must able to isolate that pathogen from the infection.

In his third and most profound postulate, he proposed that upon isolation from the pure culture of a diseased animal, the same pathogen must cause the same disease when injected into a new, healthy, unsuspecting animal. Finally, he proposed that once this pathogen causes the disease in the new animal, the pathogen must be recoverable and shown to be the same as the original.

Koch had been working as a district physician in Wollstein and doing research on the side. Propelled by his discovery surrounding anthrax, Koch ascended the academic ladder. He moved to Kaiserliches Gesundheitsamt (Imperial Health Authority) in Berlin in 1880, and in 1885 he became the first full professor at the Hygiene Institute of Friedrich-Wilhelms-Universität. The Royal Prussian Institute for Infectious Diseases opened its doors in 1891 with Koch as its director.[10] Koch remained its head until 1904. During this period, Koch's lab was among the most prolific in modern microbiology—the team made advances in science on a monthly basis and solved some of the biggest puzzles of the time, including which pathogens cause tuberculosis and cholera.

Koch's research team was composed of a series of exceptional students, interns, and visitors—including Kitasato. Kitasato was a co-discoverer of the bacteria that causes tetanus. And Julius Petri, also at Koch's lab,[11] is forever tied to the simple petri dishes used to this day in all labs doing bacterial studies. But among the most exceptional of Koch's protégés was Paul Ehrlich.[12]

Ehrlich had become a pioneer in creating highly specific and specialized dyes that would bind to specific components inside different types of cells in the blood. In 1891, Ehrlich joined Robert Koch at the institute in Berlin. He stayed with Koch for five years and then in 1896 became the director of a new institute on serum research. Ehrlich had been studying the immune system since the 1870s. During his research on the immune system in partnership with Emil von Behring, a physician working at the University of Marburg, Ehrlich wanted to know how exposure to certain microbes can create immunity in a patient. This led him to think about specific molecules released by the immune cells that can target the disease-causing microbe.[13]

Ehrlich went on to make major contributions in the field of immunology and how immune cells recognize foreign molecules and microbes.[14] But his interests went far beyond studying the immune system. He was also interested in creating specific therapies that would target microbes. While he had been focusing on the immune system, the brilliant Ehrlich had not forgotten his work on dyes, and he found a way to combine these two seemingly disparate fields. He wondered, if the dyes could bind to specific components of the cell and make them visible under the microscope, could there be other therapeutic molecules that did the same? And if so, might they shut down important parts of the cell? If this process was possible, Ehrlich argued, then specific molecules could be created to kill disease-causing pathogens. Ehrlich was correct, in both his hypothesis and his "lock and key" mechanism. He proved that a small drug molecule could enter the cell, engage with its target, and either shut down its function or kill it altogether.[15]

Ehrlich's big break came while working with syphilis, when he found that compound number 606 (later marketed as Salvarsan) attacked the pathogen but left all other cells unharmed.[16] Ehrlich's compound, in his words, was a "magic bullet"—a term that built upon the old superstitious belief that bullets could be charmed, under the right spell, to hit a particular person. His pioneering work, shaped in great part by his time in Koch's lab and subsequently at other institutes in Germany, also ushered in the era of chemotherapy—an understanding that a drug could target a specific cell while leaving other cells unperturbed.[17]

Though he was remarkably successful, Robert Koch is also known for his scientific misconduct,[18] which speaks to the imperfections of science, as well as the imperfections of those who were at the top of the scientific enterprise. The fallacies of complex characters, like Koch, were not just ethically problematic; they also caused setbacks in the progress of science. For example, Koch had announced to the world his discovery of "tuberculin," supposedly a vaccine against TB. But the vaccine wasn't successful.[19] Koch and his supporters defended the negative results by arguing that the patients were so sick that no vaccine could save them. There were other serious problems. Koch had no idea how to make a vaccine, and his formulation was harmful to those who received it, causing a serious allergic reaction. Though Koch had claimed his vaccine worked perfectly on guinea pigs, he could not produce a single guinea pig when he was asked to provide evidence of the animals he supposedly treated.

Worse was yet to come. In 1906, Koch embarked on a bold journey to cure "sleeping sickness" in German East Africa.[20] Transmitted by the bite of an infected tsetse fly, the disease is fatal if it's left untreated. The disease was affecting the health of African laborers, who were critical for the success of colonial commerce, and hence was highly relevant to European colonists in Africa.

Koch visited German East Africa and recommended using atoxyl to cure sleeping sickness. Initial results of atoxyl in animals were encouraging, but there was a problem. Atoxyl contained high levels of arsenic. The famous Koch ignored this fact and had the support of the German government to go ahead with the atoxyl treatment. Atoxyl became the drug of choice since it was cheap and stable in the tropics. But it did little to cure sleeping sickness. Instead, one in five people who took the drug became irreversibly blind. It was a disaster.[21]

Yet Koch was stubborn and refused to believe the results. He continued to believe in the efficacy of atoxyl. In the face of negative in vivo results, he was undeterred. He thought that the reason the drug was not effective in treating sleeping sickness was because the dose was too low. He recommended doubling the dose of atoxyl to be given to the native population. These recommendations were accepted by the German authorities in charge of running the clinical trials in East Africa and implemented in the region around Lake Victoria. This decision inflicted terrible suffering and sparked resentment and mistrust of German doctors among local populations.[22]

The meticulous scientist, revered by many in Germany, had few if any challengers to his authority during his lifetime; many of his fallacies came to light posthumously. The institutional protocols to question ethical use of new and high-risk therapies were simply not in place yet. Meanwhile, Koch's failings in public health have been largely forgotten. His ashes are interred in a large room with a gilded ceiling on the first floor of a museum that is connected to the institute that bears his name, an institute that is among the most respected in the field of infectious diseases.

About 650 miles southwest of the mausoleum that contains Koch's ashes is the tomb of another man whose fame, success, and worldwide impact would rival that of Koch's. From Tunis to Tehran, Shanghai to São Paulo, Bucharest to Bangui—the institutes bearing his name are a testament to his legacy.

A mile away from my own office in Boston is a street that is named after him as well. That small street is home to some of the most prominent medical research buildings in the city, and it leads straight to the marble-faced architecture of Harvard Medical School. All of it, the street and scattered institutes, pay homage to the prolific scientist and national secular saint, Louis Pasteur, a man the French citizens recently listed as the second greatest Frenchman ever, after Charles de Gaulle.[23] Not bad for a man born to a poor family in the tanning business.

Pasteur's name does not just appear on buildings—it is also printed on milk cartons sold from Delhi to Damascus. Pasteurization is standard practice in the preparation of milk, cheese, and other foods to delay their spoiling. This practical application of germ theory, by treating food with mild heat, is something we now all expect when we buy products in supermarkets throughout the world.

But pasteurization is not the breakthrough that secured Pasteur the blessings of Emperor Napoleon III. It was Pasteur's work in fermentation and wine making that truly changed how we think about the role of microbes today.[24] In many ways, Pasteur was an ideal scientist and researcher—he was someone who could both identify the problem and discover the solution.

In the early 1860s, Pasteur demonstrated that fermentation, the necessary precondition to the formation of wine itself, was dependent on the simultaneous "organization, development and multiplication" of microscopic animals. This finding was a huge achievement, not only because of the discovery's impact on an industry and product the French cared about but also because of its potential commercial impact. Pasteur's insight, and the processes that followed from it, promised to greatly extend the shelf life of innumerable food products. Soon Pasteur was a national hero—and within a decade, he was receiving about 10 percent of all funding spent on science and technology in France to conduct his research.[25] More important, Pasteur was able to show how microbes organize and how they are responsible for disease (and not environmental factors or the weak physical state of a patient).

Pasteur's star was rising as rapidly as Koch's in Germany, but this was not the reason the two men nursed a barely disguised hostility. The animosity between them could be traced back at least as far as the War of 1870 between France and Prussia.[26] But Pasteur and Koch's disputes blinded them from something they would be loath to admit—they had a lot in common. Both had concluded, independently, that germs cause disease. Both were skilled lab technicians, and both were very conscious of their status as the leading scientists in their respective countries. They both had trained rigorously and had benefited from the mentorship of other senior scientists—in Koch's case it was Cohn; in Pasteur's case it was the astronomer Jean-Baptiste Biot.[27] And both men had a tendency to allow ego and fame to cloud their judgment and behavior.

Five years after Koch wrote a paper on anthrax, Pasteur claimed that he had developed a vaccine for anthrax that worked by exposing the toxin to oxygen. To demonstrate its efficacy, Pasteur created a dramatic stunt on May 5, 1881. Working with French farmers in the village of Pouilly-le-Fort in the borough of Melun, he divided fifty sheep into two groups. He administered the vaccine to only one group, and then exposed both groups to the anthrax toxin. The results were striking. Within a month, none of the sheep that were vaccinated fell sick, and all of the control group had died of the disease.

Pasteur's public success further cemented his status as a major scientist of his time, but Koch was not impressed. Professional jealousy and fierce competition to claim the title of "greatest living bacteriologist" was one factor. National origin was another. Pasteur was a Frenchman, Koch a proud German. At one point, Pasteur even remarked: "Hatred to Prussia. Vengeance. Vengeance. Vengeance."[28]

By the mid-1880s, Pasteur had solidified his reputation as a world-famous scientist, but he wasn't done, not with scientific advances and not with greater fame, often accomplished through his carefully choreographed and publicly conducted clinical trials. In July 1885, Pasteur gave an experimental vaccine for rabies to a young boy named Joseph Meister after his mother pleaded for someone to help her sick child. Pasteur claimed he had developed this vaccine through successful experiments using fifty dogs. Inoculated, the boy survived, making Pasteur even more celebrated and his research findings unassailable. Pasteur's successful rabies vaccine also enabled him to raise even greater sums of money for his research. Riding on his fame, Pasteur pushed for an institute on vaccine development.

The Pasteur Institute lives on to this day in France and in branches around the world. It also speaks to another similarity between Pasteur and Koch. For all their undoubted genius, both men were willing to engage in unethical practices.[29] Koch tried to disguise his faulty TB vaccine through embellishment and falsified results, an act that was visible during his lifetime, but Pasteur's practices came to light decades after his death. Pasteur's will had stated that his lab notebooks should be withheld from the public. His wish was honored for eighty years. Then, in 1965, Pasteur's grandson, Dr. Pasteur Vallery-Radot, donated the notebooks to the Bibliothèque Nationale, the national library in Paris.

He did so on one condition. Access was to be limited until his *own* death. And when the collection of these detailed lab notebooks was finally opened to the public, a demigod was replaced by a more complex human being. The Pasteur who emerged was ruthless, vicious, and at times misleading.[30] His grand claim to have made the anthrax vaccine by exposing the toxin to oxygen was false. Instead, he had used the methods of one of his rivals: Jean-Joseph Toussaint.[31] Not only had Pasteur never given Toussaint any credit, his claim to having developed a unique method of producing the vaccine gave him a monopoly over it. And his publicity-grabbing claim to have saved Joseph Meister from rabies through his vaccine was also false: the vaccine had never been tested on any animal, contrary to what Pasteur had claimed. Even by the standards of the time, trial vaccine on humans—with no prior experiments on other animals—was considered unethical. The boy's recovery was more likely due to chance than to the drug.

Koch and Pasteur memorialize something often missed in the history of humankind's endless contest with disease-causing bacteria. We confront bacterial pathogens that are beyond numeration. These tiny little creatures, through endless cycles of division, adapt on us, within us, all around us, and in the most remote and inaccessible places. Following the rules of Darwinian evolution and random mutations, the ones that are able to inherit or acquire incremental advantages tend to survive and outsmart the competition. Against the threat of pathogenic bacteria that are constantly evolving, we have frequently marshaled genius, and genius is, in turn, frequently inspired by envy—national and personal—and driven by ambitions—ethical and not. That genius has often been flawed at a personal level, as was the case with both Koch and Pasteur. Yet the harnessing of genius, the organizing of its best instincts and the corralling of its worst, would be the work of institutions that would arise slowly over the remainder of the twentieth century and continue to shape our fight for survival to this day.

CHAPTER 8

FROM THE PHAGES OF HISTORY

The human body is often a host to parasites. Fleas can live off our blood, mites in our dead skin cells. Thorny-headed worms prefer our intestinal track, whereas roundworms are less discriminating, thriving in our guts, our blood, our lymphatic system. And on the microscopic scale, protozoa, infectious and not, live off our blood and tissue. In many instances, the parasitic invasion proves fatal, which is why the discovery of bacteriophage was and remains one of the more hopeful advances in humankind's contest with bacteria.

Viruses need a living host to survive—they cannot just live on the surface of a bedside table, or on the doorknob. Among the different types of viruses are bacteriophages, or phages for short. Phages are tiny viruses that infect and commandeer the bacterial machinery.[1] They can also take DNA from one bacteria and transport it to another. From a clinical point of view, phages can be useful for a simple reason—they live inside bacteria. And because of where they live, and because they can control bacterial function, they can also kill the host.

As is nearly always the case, the discovery of phages or their use as a potential treatment cannot in truth be traced to one person alone, though one man—the French Canadian biologist Félix d'Hérelle—did much to try to claim that distinction. In his account published in 1926,[2] he describes an exploration that took him around the world, from Mexico to Argentina to North Africa, and outlines how he survived World War I, then ended up in Paris. In Paris, by d'Hérelle's account, he discovered bacteriophages while caring for patients with bacillary dysentery caused by *Shigella* bacteria.[3] While working with those who had been recovering from dysentery, d'Hérelle noted the presence of what he called an anti-shiga microbe. When he filtered this anti-shiga microbe and exposed disease-causing bacteria to this filtered solution, the solution killed all the bacteria.[4] He called these tiny microbes he filtered bacteriophages, meaning bacteria-eaters. What we do know is that he saw with clarity their potential to combat bacteria. Also certain, his declared discovery and argument that these viruses could be used to treat bacterial infections, especially for ailments such as dysentery, which made him one of the reigning superstars of science in the years following the Great War.

D'Hérelle's story, told in his own words, is rich, powerful, and fascinating. It is also significantly embellished. The problem is that d'Hérelle's fascinating, globe-trotting story did not appear in the first French edition of his book, published in 1921.[5] The story appeared in full detail only in the 1926 edition. Somehow, d'Hérelle's memory had received a boost over the years. The new material was due, in part, to incessant attacks on his claim to have been the sole and original discoverer of phages. Scientists, particularly the Belgian scientists Jules Bordet and André Gratia, pointed out that d'Hérelle was not the discoverer, and that he was knowingly misleading the scientific community and, indeed, the world.[6] Citing their evidence, Bordet and Gratia declared that the true discoverer was an Englishman named Frederick Twort—who had published his paper demonstrating his finding two years before d'Hérelle.[7]

Twort had been trained as a doctor, but he turned his attention to lab research in bacteriology, and early in his career invented a new stain to dye bacteria. The dye research, which had been a major thrust of Ehrlich's research and had been pioneered in Germany, was continuing to shape other careers in bacteriology in other parts of Europe.[8] Twort's method was an improvement over Hans Gram's method of classifying bacteria, by which he would use a series of dyes and decoloring agents to see whether they would take the stain or not. Twort was a stickler for protocol, and in part out of respect for the older scientist's pioneering work the newer method was called the Gram-Twort stain. He was an unusually productive researcher and discovered right before the start of World War I a mechanism for growing and culturing a bacteria that was responsible for a wasting disease in cattle. And then, in 1915, Twort published a paper in which he reported seeing what d'Hérelle would later label phages.[9]

Twort noted that he saw dead bacteria under a microscope that were glassy and transparent. He also noticed that the glassy spots could proliferate. He suggested three possibilities: First, some previously unknown aspect of bacterial life was manifesting itself. Second, it could be an enzyme that bacteria were producing. Third, it could be "ultra-microscopic virus." This third possibility was the most audacious—it meant that viruses could infect and colonize bacteria, control their function, and, at an opportune time, even kill their hosts. Perhaps indicating Twort's own best guess, the paper was titled "An Investigation on the Nature of Ultra-Microscopic Viruses."

It wasn't just humility, however, that separated Twort from more public-savvy scientists such as d'Hérelle or Pasteur. The last line of his paper helps tell the tale of how the scientific enterprise, and research funds, supported the hyperbolic and not the humble. "I regret," he wrote, "that financial considerations have prevented me carrying these researches to a definite conclusion."[10]

Twort's career, like much of antibiotics research in general, was shaped by the Great War. During the war, Twort enlisted in the army and was stationed in Thessaloniki, Greece, where malaria was a major challenge for the population there. As a result, he couldn't continue his work on phages due to the immediate needs of taking care of the injured and the sick. Basic research on phages could not be a priority during the war.[11]

When he was pressured to give Twort due credit for his work on phages, d'Hérelle continued to maintain that he had not seen his findings. Many within the microbiological community found it hard to believe that someone like d'Hérelle would be ignorant of Twort's work. Some leading scientists insinuated that d'Hérelle, willing to steal an idea from a fellow scientist, was not a gentleman.[12] But d'Hérelle was undeterred, continuing to move ahead with his work on phages, claiming it as his right to pursue and improve what he had discovered. He saw tremendous potential for phages. Their ability to kill host bacteria held incredible promise.

D'Hérelle used his fame to conduct clinical trials and travel the world. From Punjab in India to Saigon in French Indochina,[13] d'Hérelle's phages were used to treat patients suffering from dysentery and even plague. Soon, phage therapy was being used in hospitals at a global scale. Hospital reports indicated outstanding results treating patients with a variety of illnesses.

The most flattering acknowledgment of d'Hérelle and his achievements was rather unexpected. He was utterly surprised when his work became the centerpiece of a book written by Sinclair Lewis. To be certain, *Arrowsmith* was a fictionalized account of the advances of d'Hérelle and others, but it captured the imagination of the reading public. The book was an instant hit, and for years it remained a staple for all medical students. In 1926, it won the Pulitzer Prize, cementing its reputation. That Lewis for his own reasons chose not to accept the prize brought *Arrowsmith*, if anything, even greater notoriety.[14]

D'Hérelle was reveling in the attention. He was soon appointed a professor at Yale. But he frequently clashed with the dean, who balked at d'Hérelle's unusually high travel commitments (and costs) and interest in commercial ventures. Frustrated, d'Hérelle left his position at Yale in 1933 and delved into another venture that would affect his team and shape phage research for decades.

As the United States and most of the world suffered during the Great Depression, Stalinist Russia spied an opportunity to assert not only that the end of capitalism was near but that communism in general, and the Soviet model in particular, was destined to bring peace and prosperity to all. Doing so entailed demonstrating Soviet superiority in all endeavors, including advancing science, specifically Soviet science. This included efforts, led by Soviet scientists like Trofim Lysenko, to discredit the notion of "Western" genetics and to assert a uniquely Soviet model of genetics.[15] Bacteriophages were instantly attractive to the Soviets. Here was a solution that tackled persistent infections that regularly afflicted the Russian people and the Red Army, and it also provided the Soviet authorities with evidence of a clinical intervention that did not need Western genetics to explain it.

In the early 1930s, right around the time d'Hérelle was concluding his professorship at Yale and packing his bags to return to Paris, he received a letter from a former protégé, Giorgi Eliava. Decades earlier, Eliava had worked with him at the Pasteur Institute. Since returning to the Soviet Union, Eliava, the handsome and articulate scientist from Stalin's home province of Georgia, had courted connections with the party bosses in the power corridors of the Soviet Union.[16] At a time when d'Hérelle had become a global celebrity, referred to (by no less an authority than the official Soviet newspaper, *Pravda*) as "one of the most outstanding microbiologists in Western Europe,"[17] Eliava could boast direct contact. And with official sanction, he issued an invitation for d'Hérelle to pursue his research in the USSR.

Eliava had established the Tbilisi Institute of Microbiology, Epidemiology, and Bacteriophage in 1923. He asked d'Hérelle if he would be interested in solidifying the reputation of the new institute. D'Hérelle agreed immediately and arrived with his wife in October 1933; he visited the institute off and on until May 1935.[18]

He brought with him his international star power, which was very much courted by the Communist government, as well as his laboratory equipment from Paris. The head of the People's Commissariat of Health offered d'Hérelle the opportunity to head any institute of his choosing in Moscow, but d'Hérelle declined. He wanted to stay in Georgia to honor his commitment to Eliava (and because the weather there was much more pleasant than in Moscow).

Nonetheless, d'Hérelle was impressed by what he saw in the USSR and was grateful to his host country. He completed work on a book while in Tbilisi and dedicated it to Stalin, doing so in the most fawning of terms. In a similar vein, he would declare the USSR "a remarkable country, which, for the first time in the history of mankind, chose as its guide not irrational mysticism, but a sober science without which there cannot be any logic or genuine progress."[19]

Shortly after d'Hérelle made those remarks, the security apparatus of the Soviet Union would kill his friend and protégé. In May 1935, the d'Hérelles left Georgia, expecting to return again as soon as they could. But the Stalinist purges had begun, and in January 1937, Eliava and his wife, Amelia Wohl-Lewicka, were arrested as enemies of the state. The tragic, and ironic, charge was espionage for France. The scientist's connection to the Pasteur Institute, which had brought Eliava and d'Hérelle together, was cited in the trumped-up accusation. In a final absurdist twist, Eliava was found to have been poisoning wells with his bacteriophages, the very scientific advance that the Soviet papers, just years earlier, had extolled. Eliava was shot dead on July 10, 1937, and his wife soon thereafter.[20]

For a few brief years in the 1930s and early '40s, the Soviet Union had been at the forefront of research into the potential for phage research to cure an array of infections. And while Eliava's Institute itself survived the purges of its senior staff, research in bacteriophages in the Soviet Union went into dormancy for decades. D'Hérelle's lifetime work was soon overshadowed by a new class of drugs: antibiotics. Little did people know that nearly eighty years later, in the early 2000s,[21] phages would become relevant again when war, greed, and bad policies would render antibiotics impotent.

CHAPTER 9

SULFA AND THE WAR

Waldemar Kaempffert, the irrepressible science editor of the *New York Times*, was skilled at keeping track of the scientists of his era. Many of the most notable were eccentrics, most of whom responded well to Kaempffert's flattery. On June 12, 1931, for example, on the occasion of Nikola Tesla's seventy-fifth birthday, he wrote a letter to Mr. Tesla. It read, in part,

> Dear Mr. Tesla,
> As I look back over thirty years of editorial and journalistic work confined to the interpretation of science and engineering your figure looms larger than any with which I have been brought in contact. It was a privilege to bring to the public notice the results of your epoch-making experiments in electrical engineering and electrical resonance."[1]

Tesla's oddities, along with his achievements, drew public attention. Other scientists were very different. In 1950, Kaempffert wrote to the editor of the *Journal of the History of Medicine and Allied Sciences* about a nearly forgotten man: "Many of those interested in chemotherapy may have wondered what had become of Dr. Paul Gelmo, who discovered sulfanilamide in 1906, only to bury his results so successfully that they had to be rediscovered by Dr. Gerhard Domagk of the Interessengemeinschaft."[2]

In 1908, Paul Gelmo discovered a synthetic antibiotic, sulfanil-amide. Gelmo, an Austrian chemist, had the wherewithal to pat-ent his finding in 1909, even though he did not fully understand how it worked. And then he did nothing with it.[3]

For two decades no one appreciated the potential power of sul-fanilamide as an antibacterial. Sulfanilamide became a block-buster in the early 1930s,[4] a drug that could rapidly treat stubborn infections in children and adults and decrease the rates of hospi-talization. The accolades and fame associated with sulfanilamide, recognized as the first commercially available antibiotic, would go to the German bacteriologist Gerhard Domagk.

Domagk was a veteran of World War I and a physician turned scientist. After the war he joined the pharmaceutical industry, and while working for Bayer in the company's laboratories in Wuppertal, Germany, he rediscovered Gelmo's original finding. Taking the rigorous route most German chemists took at that time to identify potential drugs, Domagk used hundreds of dye mole-cules, most of them variants of existing chemicals, to figure out if they were effective against *Streptococcus species*—these bacteria were a good model system for infection as they would reliably kill the lab mice through sepsis. Nothing worked, and he was left with hundreds of dead mice that did not respond to any molecule that he hoped would have curative effects.[5]

In 1932, Domagk and his team tried using their dyes with sulfa-nilamide on mice that he had infected with strep. Miraculously, it worked against the strep infections in mice. Over the next two years, scientists at Bayer showed that it worked on other infec-tions as well, including pneumonia, spinal meningitis, and gon-orrhea. Bayer called the drug Prontosil.[6]

For Domagk, the impact was beyond global—it was also deeply personal. In December 1935, his six-year-old daughter Hilde-garde developed an infected abscess in her hand. The infection spread and her temperature reached 104˚F. Blood tests showed she had contracted a severe streptococcal infection. Hildegarde moved in and out of consciousness, as her life hung in the balance. Had she been infected a year or two earlier, or been born somewhere else in the world, she would have died.[7] But she survived—a week after getting Prontosil from her father, she was fully recovered, playing again in the yard.

When Bayer found out that this blockbuster drug was not an original drug, but had already been discovered in Austria in 1909, it was a distressing blow to the company.[8] Their claim to exclusivity was severely limited, and competitors soon introduced their own sulfa-based synthetic antibiotics. Bayer did its best to rely on the company's prowess with marketing and branding, and for a time Prontosil album, the company's full name for the drug, was successful. The wonders of the drug continued to capture headlines around the globe as it was used to treat life-threatening ailments, and competitors continued to come to market.

Different companies introduced the drug with their own names and brands. Thousands of lives were saved by the drug, including that of a young man whose father was the president of the United States. Franklin Delano Roosevelt Jr. had developed a strep infection in December 1936. A decade before, the outcome would have been different. But he was successfully treated with the drug— trade name Prontylin. On December 17, 1936, the front page of the *New York Times* read:

YOUNG ROOSEVELT SAVED BY NEW DRUG
Doctor Used Prontylin in Fight on Streptococcus Infection of the Throat.

CONDITION ONCE SERIOUS But Youth, in Boston Hospital, Gains Steadily—Fiancée, Reassured, Leaves Bedside.

The fact that the drug was off-patent meant that it was easily available. And its tremendous potency against strep throat meant that demand skyrocketed.[9] To carve out some of that demand meant that companies sought formulations that would differentiate their drug from the rest. For example, since sulfanilamide (sold as Prontosil and Prontylin) was not water soluble, it was difficult for children to swallow. Spying an opportunity, Harold

Watkins, chief chemist at S. E. Massengill Company, based in Bristol, Tennessee, came up with a new formulation. He dissolved the powder in a chemical called diethylene glycol and even added raspberry flavoring. The company tested the appearance, fragrance, and flavor of the drug, and then shipped a total of 240 gallons of it to doctors across the country.[10]

But the drug did not cure strep throat—instead, it resulted in a painful death due to renal failure. It was a national tragedy. Doctors from across the United States started reporting the horrid outcomes. After he endured the death of six of his patients, one of them his best friend, due to a drug he administered, Dr. A. S. Calhoun wrote, "I have known hours when death for me would be a welcome relief from this agony."[11]

In 1906, President Theodore Roosevelt signed the Pure Food and Drug Act. Among other things, the passage of the act was made possible by the work of a group of volunteers called the Poison Squad, the brainchild of an Indiana scientist by the name of Harvey Wiley. As the head of the bureau of chemistry in the US Department of Agriculture, Wiley had been taking a keen interest in adulterated and poisoned food available in the marketplace. To test the food for safety, he decided to create a group of volunteers whose members were to test on themselves certain foods for toxicity, at times at the cost of their own health.[12]

Wiley was not just bold in his efforts—he was also persuasive and well connected. He convinced President Theodore Roosevelt that there had to be an act that protected consumers from harmful food and drugs. The 1906 Pure Food and Drug Act was a result of his political savvy. By 1927, the small office of the bureau of chemistry in the Department of Agriculture had become its own institution—the Food, Drug and Insecticide Administration.

Now, as the sulfanilamide tragedy was unfolding, another Roosevelt was in the Oval Office. Under his watch, the FDA sprang into action and recalled whatever they could of the liquid form of sulfanilamide from all over the country. An analysis by the government of why the problem originated showed that there was a clear loophole in the policies of the time. The law expected drug companies to report only drug efficacy. There was no requirement for the pharmaceutical companies to conduct or report any toxicity tests on drugs coming to market. This loophole was finally closed, with the passage of laws that required extensive testing of drugs for toxicity and sharing of those results with the FDA. The role of the FDA changed forever, which proved to be of critical importance as new drugs were coming to the market. They had to be effective and safe.[13]

The same sulfa drug that saved FDR's son had become a routine and excessively used therapy for infection and injury by the time US forces entered World War II. Like the brave men who risked their lives, sulfa drugs were a key part of the arsenal at the country's disposal. Colonel Elliott Cutler, a man with sharp insight and hostility toward politicians, including FDR, was the chief consultant of surgery in the European Theater of World War II. He was overseeing the mass administration of the drug in the battlefield and in the field hospitals at the time.[14]

In 1943, Cutler observed something quite disturbing. American troops at that time were primarily engaged in fighting on the African continent. Tasked with overseeing the care of all US servicemen fighting in Europe, Cutler initiated a study of the wounded soldiers returning from North Africa who had been given sulfa drugs. The results from the 332 cases were shocking. Cutler concluded in May 1943, "The statistics show that the sulfonamides, even taken and given under the optimum conditions, do not keep infection away from wounds."[15]

The drug that had been considered a miracle cure during the last decade was now failing, and not because of manufacturing problems, or because the nature of the infection had changed. The bacteria were now winning the battle. They had figured out how to evade the assault and had become resistant to sulfa drugs. Cutler realized that while it may have worked wonders a decade earlier, it was no longer effective.

Cutler was later asked by one of the members of the British Parliament: "Can it be said that the Sulfonamides as used by the US Armed Forces have saved life?"

Cutler responded: "The answer must be no."[16]

The US Army had not just used the miracle drug—it had come to know the limits of that miracle. Resistance, now, was just as real as the drug. But Cutler knew of something else—a phenomenon that would later influence the practice of army doctors and physicians around the world. He knew the power of believing in a drug, even when it didn't work. As he wrote: "Even transcending the above deductions of the importance are the psychological effects upon the troops themselves. Almost to a man the soldiers have said, when questioned, that their lives were saved by the use of Sulfa drugs. Experienced clinicians will recognize the values of this mental attitude, and whether recognized or not by the physical scientists of this generation, it is something no good physician would be willing to set aside as a highly beneficial agent in the recovery from any physical ill."[17]

It is unlikely that Cutler knew that overuse of the drug in the battlefield was responsible for bacterial resistance, but he clearly knew that even when the drug did not work, there was a reason to prescribe it. It had more to do with the collective faith in science and less to do with the drug's efficacy.

This belief factor continues today. Patients demand—and expect—that they will be given an antibiotic when they have a fever, or when the patients perceive that they have an infection. Doctors around the world often prescribe antibiotics, knowing that the power of the drug is limited and is rapidly decreasing. But these drugs can be a source of hope for patients, and, like Cutler during the war, many doctors believe in the power of perception and the faith patients have in the drugs—even when they do not work.

The problem is not with demand alone. Many doctors are also not fully aware of the broader impact of overprescribing antibiotics and are eager to write a prescription for drugs that have worked in the past, and are, for the most part, quite easily available. Another version of this ethical conflict is played out in primary care centers all around the world. Patients may believe that they need antibiotics to get better, and when doctors are unwilling to prescribe them, the patients go to other doctors who are willing. In a world increasingly concerned about ratings and patient reviews, doctors who are worried about the growth or sustainability of their practice choose to go with writing a prescription for antibiotics, the efficacy of which may be questionable, rather than deliver a long sermon about why patients don't need them.

Cutler was not just taking care of his soldiers. He was also involved in a US-supported clandestine program that took him to Moscow at the height of the war. The goal of the program was to provide Stalin, and his army, with a precious new drug called penicillin.[18] Little did Cutler know that the same errors of judgment that had rendered sulfa drugs ineffective would make the latest miracle impotent as well.

CHAPTER 10

MOLD JUICE

On December 11, 1945, a slender, clean-shaven Scotsman was taking the stage in Stockholm. The day before he had met the tall monarch of Sweden, King Gustav, who towered over the scientist. Now Sir Alexander Fleming was waiting to be awarded the Nobel Prize for his work on the discovery of penicillin.

When Fleming gave his acceptance speech, he was so soft spoken that the attendees strained to hear him. Toward the end of his speech, the Scotsman sounded a word of warning about the limits of the very discovery for which he was being celebrated:

Penicillin is to all intents and purposes non-poisonous so there is no need to worry about giving an overdose and poisoning the patient. There may be a danger, though, in underdosage. It is not difficult to make microbes resistant to penicillin in the laboratory by exposing them to concentrations not sufficient to kill them, and the same thing has occasionally happened in the body. The time may come when penicillin can be bought by anyone in the shops. Then there is the danger that the ignorant man may easily underdose himself and by exposing his microbes to non-lethal quantities of the drug make them resistant. Here is a hypothetical illustration. Mr. X. has a sore throat. He buys some penicillin and gives himself, not enough to kill the streptococci but enough to educate them to resist penicillin. He then infects

his wife. Mrs. X gets pneumonia and is treated with penicillin.
As the streptococci are now resistant to penicillin the treatment
fails. Mrs. X dies. Who is primarily responsible for Mrs. X's
death? Why Mr. X whose negligent use of penicillin changed the
nature of the microbe.[1]

It was a poignant question being raised by a man who was a hero
for millions. Fleming's discovery of penicillin had not only helped
the Allied Powers win the war, it had also helped sick people the
world over. And after posing the question of what happens when
doctors are not cautious, Sir Alexander Fleming provided an answer. Mr. X killed his wife through his "negligent use of penicillin." The moral was clear, and Fleming was unequivocal. "If you
use penicillin, use enough."[2]

The story of Fleming's discovery of penicillin is often recounted
as a way to emphasize the concept of serendipity. It runs like
this: in August 1928, Fleming was in a rush to go on vacation. In
his haste, he later claimed, he had left a window open in his lab.
Close to the window were petri dishes with cultures of *Staphylococcus aureus* bacteria. When Fleming returned from vacation
in early September, he discovered the open window and the petri dishes, all of which looked fine except for one. The lone dish
stood out because of the obvious fungal contamination, and the
contamination had created a ring on the plate. All of the bacteria
that had come in contact with the fungus had died. Fleming concluded that the fungus—which he described as mold called *Penicillium notatum* in a later study—had something in it that could
kill bacteria. His "mould juice,"[3] he concluded, had the potential
to be a remarkably effective bacteria killer.

The story has been told countless times to teach young scien-
tists that chance and luck are integral to their chosen profession.
Except the story isn't entirely true.[4] It's very similar to the one
Fleming told concerning another discovery of his: lysozyme, an

antibacterial enzyme present in a large number of biological substances, from nasal mucus to egg white. And it strains credulity
to imagine that a forgotten open window could play so central a
role in the discovery of two antibiotics. Those familiar with the
history of science and medicine at the time have concluded that
Fleming's story was embellished in part because he liked to tell
fantastic stories, and because a true story probably suggested that
he was at times forgetful, sloppy, and perhaps not as rigorous and
careful a scientist as he claimed to be.[5]

Without doubt, when Fleming saw the ring of fungus, he recognized that there was something potent in the dish. His mold
juice would occupy significant amounts of his time, as well as
that of his trusted associate Stuart Craddock, as they tried to understand its role in killing bacteria. The mold juice had lots of
impurities, and the concentration of the effective agent, penicillin, needed to kill bacteria was no more than 1 percent. Fleming
was not a chemist, and the results he was seeing in his lab were
plagued by his ineffective distillation techniques and the consequent poor overall efficacy of the mold juice. Since he was not
making progress on purifying penicillin, by the mid-1930s Fleming had moved on, or rather moved back, to his first discovery
of lysozyme. So had the world: the wide use of sulfa drugs at the
time meant that Fleming's original discovery, published in a paper in 1929, was largely forgotten for nearly a decade.[6]

Toward the end of the 1930s, Fleming's penicillin paper gained
attention from a team of eclectic and often mutually distrustful
researchers about sixty miles west of St. Mary's Hospital, where
Fleming had made his discovery. Based in Oxford, the team was
part of the Sir William Dunn School of Pathology, and among its
researchers was Howard Florey, an Australian Rhodes Scholar
pathologist and the newly minted director of the Dunn School.
By his side was a brilliant chemist, Ernst Boris Chain. A Jewish
refugee from Germany, Chain was supported by a progressive
group called the London Jewish Refugees Committee. The third

member of the team was Norman Heatley, who was trained as a biochemist but was exceptionally gifted at designing laboratory instruments.[7]

Exactly how Fleming's paper came to the team's attention is its own mystery. It might have been for no more remarkable reason than that the team members were already working on lysozyme, which would have had them reviewing all of Fleming's work. There is another theory that Cecil George Paine, a pathologist at the University of Sheffield and former student of Fleming's, told Florey about penicillin.[8] Florey was a professor at Sheffield from 1932 to 1935, where Paine had already developed an interest in penicillin, and there is evidence that he may have used crude penicillin to treat eye infections in young children.

Florey moved to Oxford in 1935, and by this time the Dunn School had established its own interest in lysozyme, as well as other natural compounds that had antibacterial properties. The team had figured out that it was difficult to stabilize penicillin and to extract it in quantities that were useful. Chain felt that the structure of the molecule had little to do with the poor yield of penicillin. The problem, according to Chain, had an altogether different origin that, if he was proved correct, was actually not hard to fix. Chain was convinced the problem lay with the incompetence of British chemists. Florey challenged Chain to prove his claim by doing what others, particularly the British chemists, had not. Never shy, never intimidated, Chain took on the challenge, wisely recruiting Heatley's help.[9]

Heatley did what neither Chain nor Florey had the competence to do. He devised instruments, with materials that were both cheap and widely available, that could help them conduct the painstaking experiments necessary to increase the purity of penicillin. Over Florey's objections, Heatley followed a hunch that he could use ether to extract the liquids from the mold, and he was eventually proved right.[10] The process was successful, improving the yield and producing penicillin in quantities that could be used in animal experiments. A process marked by boasts, challenges, ingenuity, and a determined willingness to play out a hunch had worked. Heatley passed the purified penicillin to Florey and Chain; it was now time to see whether it was effective.[11]

On May 25, 1940, Florey started his experiment with eight mice divided into two groups, four in the control, and four that were given the extracted penicillin. All eight were infected with *Streptococcus pyogenes*. The four control group mice promptly died. The other four survived. Heatley, who had been in the lab until 3:45 in the morning watching the experiment unfold, wrote in his diary, "It really looks as if P. [penicillin] may be of practical importance." That was an understatement.[12]

The effort in Oxford continued as the Great War raged on and the number of wounded soldiers increased. The publication of the results with the mice gained the team attention around the world, and the men started fielding requests for their precious drug.[13] Everyone had taken notice of what the Dunn School had accomplished, and on September 2, 1940, the team was surprised to find a middle-aged man with a Scottish accent visiting their lab. It was Alexander Fleming, who asked Chain what had happened to his "old penicillin." Fleming was likely intrigued about the research at the Dunn School as well as concerned about getting due credit for his discovery.[14]

To prove that penicillin really had a therapeutic value, the Oxford team needed to move quickly to show results in humans. There was a lot of academic interest in the new drug, but there was no commercial interest from the UK pharmaceutical companies. Too many unknowns about the drug, its potency, and its purity kept the companies at bay. To attract support and financing, the team needed to produce larger quantities of penicillin, which meant having more effective equipment that would operate at a scale they had not attempted before. Heatley, ever an ingenious innovator, designed yet more efficient apparatus that produced enough penicillin that was pure enough to give to an adult. It was time to show the world how good penicillin really was.

The first real test of penicillin's ability to cure infection in humans involved an Oxford policeman named Albert Alexander. Alexander had developed an infection that was oozing yellow pus. Despite aggressive use of sulfa drugs, over the previous few weeks the infection had moved to his lungs. He was given the newly purified penicillin drug on February 12, 1941.[15] The results were spectacular. Alexander's face had been swollen and his wound deeply infected, but the intravenous injection took care of that within a day. His fever disappeared and his condition started to improve. While not fully recovered, he was able to sit up and even eat—something he could not have done just a day before the injection.

But there was a problem. Alexander was a fully grown adult man and required lots and lots of penicillin to recover fully. And what Heatley and his team had made was far from pure. At 5 percent purity, the amount needed for Alexander was more than what the Oxford team had in all of its stock. Even as Heatley and the team raced to produce more, Albert Alexander died on March 15, 1941. Alexander's death was a blow to everyone who had been hoping for a miracle cure. Yet there was still reason to be optimistic. The drug had worked—there just wasn't enough of it. It was no longer a clinical problem that needed to be solved, but a *production* problem.

Whether the drug could be produced in mass quantities using the current method of production remained an open question. The Oxford team calculated that they would need *kilograms* of the drug, not the mere grams that they were able to produce, and of a purity greater than what they had been achieving. Purification required hard cash and investment from the government or the private sector.

By March 1941, Britain stood alone against the full force of the Nazi's Third Reich. And given the war's devastating impact on the UK, there were no English companies or institutions capable of helping the team take on the challenge. They decided to look westward. Heatley and Florey, with the help of the Rockefeller Foundation, which had been supporting Florey for over a decade, reached LaGuardia Airport on July 2, 1941, to continue their work on penicillin with potential US partners.[16]

Florey first headed to New Haven to meet his friend John Fulton, a professor of physiology at Yale. Fulton had first met Florey when they were both Rhodes Scholars in the early 1920s. They became good friends, to the point that during the war, Florey sent his children to America under Fulton's care.[17]

Now Fulton did Florey another favor. His contacts got Florey introduced to the senior leadership of the National Research Council, which was tasked with leveraging science for national security. These connections led Florey and Heatley to Peoria, a suburb of Chicago, where they were going to meet with researchers at the Northern Regional Research Labs of the US Department of Agriculture.

The lab there had already started working on increasing quantities and purity of penicillin and had sent requests all over the world in hopes of locating the best source of the mold *Penicillium*. The best sample, it turned out, came not from far-off lands but from a farmers market in a neighboring suburb. Mary Hunt, a bacteriologist working at the Northern Lab, bought a rotten cantaloupe, which ended up producing the best strain.[18] Other scientists at the lab were tasked with finding the ideal conditions to grow the mold, and creating apparatuses to increase the yield.

Unlike the Dunn Lab, where experiments were conducted under a small budget, the Northern Lab had far greater resources. The quantities that the American lab could produce were significantly greater than what the Dunn School could ever imagine. But this was still not enough for the demand in the battlefields across Europe.

In August 1941, Florey left Heatley in Peoria and came east to Philadelphia to meet a former colleague named Alfred Newton Richards. Richards was now the head of the powerful Committee on Medical Research (CMR) at the US Office of Scientific Research and Development (OSRD).[19] OSRD had been created by an Executive Order of President Roosevelt and was guaranteed resources needed for any "research on scientific and medical problems relating to national defense."[20] Hearing Florey out, Richards committed to recommending government support for penicillin production.

Even after Florey returned to the UK in September 1941, Richards continued to lobby for funding for the penicillin project. Through his help, and that of his boss Vannevar Bush, a meeting of OSRD and private sector leaders was arranged in October 1941. The meeting included members of CMR and OSRD, as well as leading pharmaceutical executives from Pfizer, Merck, and Lederle. Though penicillin was at the top of the agenda, no final decision was taken regarding funding.[21] The next meeting was to convene in December. By then, the US engagement in the war had completely changed. The Imperial Japanese Navy launched its surprise attack on the US Navy base at Pearl Harbor on December 7, 1941, and America formally joined the war.

The meeting organized by OSRD happened ten days after the Japanese attack. Now, the goal was no longer to aid the Dunn Lab but to protect all Allied troops from infection-related deaths. The best way to do that was through the prompt production of vast quantities of high purity penicillin. Through the support of the US government, the interest of large pharmaceutical companies, and the resources of the Department of Agriculture, the center of gravity for penicillin production had moved westward from England to the United States. Patents were issued in ever increasing numbers, and Chain, Heatley, and Florey were left out as the patents focused on fermentation methods and purification processes, not on the product itself.[22] More troubling was the fact that Andrew Moyer, a microbiologist and member of the Northern Regional Research Laboratory, received a patent for the development of fermentation methods for penicillin production. Though Norman Heatley worked closely with Moyer in Peoria on the process, his name did not appear anywhere on the patent.

The combined support from the US government, the expertise from the USDA, and the investment of pharmaceutical companies meant that production reached a pace previously unimaginable. Millions of dollars flowed to universities to support laboratories and tens of millions were spent by pharmaceutical companies on research and production. Sixteen new penicillin plants were approved by the US government. Massive tax breaks made it possible for drug companies to invest large sums at little risk of loss.[23]

Given the level of investment and expertise available in the United States, and the fact that its industry was not under threat of imminent attacks, progress in the United States quickly dwarfed efforts in the UK. By 1944, the United States was producing forty times more penicillin than the UK. The drug not only changed the course of the war—it changed the future of the US pharmaceutical industry as well.

The drug was considered a savior for millions. Clinical trials and use on the battlefield had demonstrated its high efficacy. But from the outset there were concerns about abuse and overuse. Fleming, the discoverer of the mold juice that had set this revolution in motion, eloquently raised his doubts while receiving his Nobel Prize. But just as the miracle drug was starting to be questioned, the Soviets, on the cusp of the Cold War, claimed that it had, in fact, been a Soviet scientist who had discovered it.

CHAPTER 11

TABLETS FROM TEARS

Fiction is frequently more convenient than facts. Consider Tatyana Vlasenkova, the fearless microbiologist beloved by Russians since the Soviet era. She is the heroine of *The Open Book*, a novel by Veniamin Aleksandrovich Kaverin that was first published in 1940. Her exploits filled a trilogy of books, became the basis of a popular television show, and were immortalized in a feature film. She is a model Soviet citizen, a gifted scientist who overcomes early adversity to perform wonders. Hardworking, a good wife and mother, she is also, of course, ready to tackle the biggest problems facing the Soviet motherland. Tatyana, the novel asserts, is the true discoverer of Soviet penicillin. Perhaps the reason Tatyana became an everyday heroine was because she was based on an actual person very close to the author—his sister-in-law, Zinaida Ermolieva.

I first encountered the name Zinaida Ermolieva in the World Health Organization's archives, in a letter written by Marcolino Gomes Candau, the second director-general of the WHO.[1] On June 26, 1959, he wrote to the Minister of Public Health of the USSR that the WHO was considering Dr. Z. Ermolieva as a potential member of the organization's advisory panel on antibiotics. She was the only woman being so considered. Upon receiving a favorable reply from the minister, Dr. Candau wrote to Ermolieva directly on August 24, 1959. She responded that she accepted the position with pleasure and would arrive in Geneva on October 5, 1959. From that point until her death in 1974, Ermolieva remained a member of the WHO's committee on antibiotics.[2]

Facts can, of course, inspire in ways fiction cannot. Though Ermolieva is one of the most famous scientists of the Soviet era, she remains largely ignored in the history of antibiotic discovery. This is unfortunate, for her life should encourage generations of scientists to come. Born in the city of Frolovo in 1898, as a young student she excelled in Latin, French, and German. Her knowledge of Latin served her especially well during her entrance exam to medical school.[3]

World War I changed her life. As the war broke out in Europe, the University of Warsaw relocated—the entire university, with all its faculties, moved to Rostov-on-Don, where Ermolieva lived. A world-class university suddenly at her doorstep, she was also fortunate to be a student when the Russian Duma voted to open medical training to women. Zinaida Ermolieva, and her lifelong friend Nina Kliueva, were part of the first class.

The years that followed Ermolieva's education were turbulent. Wars with foreign powers and domestic enemies shaped Ermolieva's career. She saw the horrors of World War I, the Russian Revolution, the subsequent civil war between the communists and the monarchists, and the famine and cholera epidemic in the immediate aftermath of the civil war. It was during these chaotic times that Ermolieva secured her first taste of bacteriology research.

She promptly made her mark. Her first paper was published in 1922. She was only twenty-three years old. And she was among the first Russian scientists to tell the difference between a cholera pathogen and a cholera-like pathogen. In a dramatic demonstration worthy of a novel's heroine, Ermolieva drank a bottle containing a cholera-like pathogen to show that a distinction had to be made between the two pathogens. She survived, her diagnosis had been correct, and her reputation was further established.

Zinaida Ermolieva rose in prominence. By the time she was

twenty-seven, she was a well-respected researcher, one who could work both in the lab and in the field. In keeping with the scientific interests of the 1920s, her early work focused on bacteriophages, the bacteria-killing viruses that were being used around the world for treating a variety of infections, and had consequently made Félix d'Hérelle famous. Her work also helped those suffering from the 1939 cholera epidemic in Iran and Afghanistan.

Her biggest break, however, came in 1942 when, under orders from Hitler, the German army had nearly encircled Stalingrad. She was transferred from Moscow to the besieged city overnight. The situation there was rapidly deteriorating. The water lines were contaminated, threatening a devastating cholera epidemic that would accomplish what the Nazis had so far failed to do: force the city to capitulate.

Her team, on instruction of the Soviet Commissariat on Health, created a secret underground lab in Stalingrad.[4] There she conducted research, created prevention measures, formed treatment clinics, and pushed for the chlorination of the water. At the height of the war, nearly fifty thousand people were treated by her phage therapy for diarrheal diseases every day.

At the end of 1942, Ermolieva received a phone call. On the other line was a man with a thick Georgian accent, whom she recognized immediately. It was the general secretary of the Central Committee, Joseph Stalin. He asked her a short question: Was it safe to keep a million people in Stalingrad, given the possibility of a cholera outbreak? She replied confidently that everything was under control—that she had won her battle. It was up to the Red Army to win theirs. For her work, Ermolieva was awarded the Stalin Prize, which, in true patriotic fashion, she donated to the war effort.[5]

Though Stalingrad had been saved, the war was not yet won. Soviet leadership was now appreciating the potential of penicillin to treat its soldiers and save its citizens. Aware of Fleming's original discovery and the advances of Florey, Chain, and Heatley, the

Soviets were making their own determined effort. Despite being an ally in the war, the Soviets were treated with suspicion by the United States and England, a suspicion they returned in kind. To win the war and demonstrate to the world the superiority of the Soviet system, they needed to reliably produce penicillin.

Ermolieva was assigned the task, and she almost lived up to the expectations of making a Soviet penicillin. She discovered *Penicillium crustosum*, a species separate from that used by the British and the Americans, and this was fast-tracked to clinical trials by early 1943. (In 1944, Florey visited Russia and met with Ermolieva.)[6] The Soviets told Florey that theirs was the most potent form of the drug yet, a fact hailed as another sign of Soviet success, such that on March 17, 1948, *Pravda* declared penicillin a Soviet discovery. But there was little actual data to back the claims of potency made by the Soviet state. Unfazed by the Soviet propaganda, Florey saw through the hype and was largely unimpressed by the production methods and the results.

Over time, as the research failed to deliver the desired results of drug efficacy, the state started to lose its patience. The fate of many of the Soviet scientists who worked on expanding the penicillin effort remains uncertain. But we know this: Vil Zeifman, who was in charge of expanding the Soviet effort, was arrested by the Stalin regime, interrogated, and exiled to Siberia.[7]

While Ermolieva's star continued to rise, her personal life was marred by difficulties and an all-too-familiar tragedy within the Soviet Union. She was able to save her first husband, Lev Zilber, from Stalinist purges several times. Despite being a well-known oncologist, he was repeatedly arrested, and sent to labor camps, for espionage. Ermolieva personally intervened in 1930, 1937, and 1940 to save his life.[8] Her second husband, Alexey Alexandrovich Zakharov, was also a scientist, and he, too, was denounced during the Stalinist purges and died in a prison.

Just as the fictions demanded by Soviet politics had robbed the country of many of its most talented scientists, they also belied the extent of their accomplishments. Eventually, it became clear that the purity of their penicillin was highly questionable and its efficacy uncertain. Ultimately, the Soviet claims of original discovery, and their accusations of an imperial plot to thwart Soviet science, did not pan out. To save the lives of its citizens, the Soviets had to buy a license for penicillin from the Europeans.[9]

CHAPTER 12

THE NEW PANDEMIC

Bacteria cares not at all for the politics of nations or the egos of scientists, and in its unrelenting drive to endure, it cares not at all for the timetables of human beings. When Fleming gave his Nobel Prize lecture in Stockholm, he could not have known that his prophecy would come true so soon. Within a year of his address, a brilliant bacteriologist named Dr. Mary Barber would sound the first alarm announcing a new pandemic. It began in Hammersmith Hospital in London.[1]

During World War II, Barber worked on the challenge of cross infections, or how patients infect each other. Hammersmith Hospital was suffering an outbreak of streptococcal sepsis. Like her peers throughout the United Kingdom, Barber was using penicillin to treat infections, and like her peers, she was pushing the limits of the drug.

In 1946, Barber started noticing something she hadn't seen during the war years. She worked from samples taken from patients suffering from various infections and saw that they were no longer responding to penicillin. As she investigated, she found that the problem was much worse than she thought. Not only was the resistance real, it was increasing. Her data was startling, puzzling, and troubling: out of one hundred cases of infection by *Staphylococcus pyogenes*, thirty-eight were penicillin resistant.[2]

Renowned as a careful researcher with exacting standards,

Barber analyzed and reanalyzed her data. Her conclusion was bold and ahead of its time. In a 1947 paper she wrote, "It is obvious that the main cause for this increase in penicillin-resistant strains of S. pyogenes is the widespread use of penicillin."[3] Fleming's warning had been loudly repeated by Mary Barber. Now the world had to respond with appropriate alarm.

The Public Health Lab (PHL) in the United Kingdom was founded in 1946 due to concerns within the country about manmade threats—biological warfare, to be precise. Discussions regarding a permanent entity that would be dedicated to protecting the public from epidemics due to wars and warfare began in the mid-1930s. England had depended on the research efforts of universities up until that time, but the government wanted to put this new institution under its own authority. Ultimately, the Ministry of Health put the PHL under the supervision of the Medical Research Council (MRC), and it finally became a reality.

In the postwar years, the main task of the lab was to provide free epidemiological services to the entire country and to the communities in the British Commonwealth. The design of the lab was such that there would be a central node, in Colindale in North West London,[4] and a network of regional sites in Oxford, Cambridge, Cardiff, and Newcastle, a plan that would eventually expand to include twenty-five smaller regional labs.[5]

Soon after its inception, the lab in Colindale became the center for research into infectious diseases. It also assumed responsibilities for mobilizing global efforts to identify, understand, and tackle epidemics, often originating well beyond British shores.[6] By the mid-1950s, about a decade after its start, nearly a thousand people worked in Colindale, and researchers from around the world communicated with and sought the aid of the staff there. Many scientists sent their lab results to Colindale, asking that the strains of a particular disease be compared against their stored samples. While working there, the researcher Phyllis Rountree set in motion a chain of events that would help define modern hospital protocols of hygiene in the face of infection.[7]

Rountree was far smarter than most of the boys she knew when she was growing up, including those in her own well-educated family of doctors, nurses, and pharmacists. She entered her undergraduate class at sixteen—a remarkable feat in 1927 Australia, which did not prevent her first boss from telling her: "It was very nice having you here dear, but we don't employ women permanently"[8] (and despite his lab being short on technically competent staff).

It was Sir Frank MacFarlane Burnet, whose name now graces a prestigious institute in Melbourne, who gave Rountree her first real job. She trained in Australia, subsequently worked in London, and then helped with the war effort, all of which sharpened her skills. By 1950, when she was awarded her PhD from the University of Melbourne, Rountree was considered an expert in bacteriophages and was working at the Prince Alfred Hospital in Sydney.[9]

In 1952, at the Royal North Shore Hospital in Sydney, doctors started observing an unusual infection in newborn babies. The staph infection was also traveling from the breastfed babies to their mothers.[10] The infection was one problem, but it was not the most concerning: it was by this time being treated with an aggressive penicillin regimen, and it was unresponsive to the wonder drug.

The doctors on call in the pediatric ward included Clair Isbister, the hospital's chief pediatrician. She would later become Australia's most famous pediatrician, in part due to the extraordinary popularity of her radio show, the "Woman Doctor of the Air."[11] In 1952, Isbister, and her colleague Beatrix Durie confronted a crisis and sought out Rountree's help. Staph infections in the ward were becoming increasingly unresponsive to their routine therapy of penicillin. Why, they desperately needed to know, were their drugs useless against these infections? Rountree got to work. She would soon find out that she was seeing a new form of staph infection that heralded one of the first global epidemics of antibiotic resistance.

Rountree used phages to identify the type of bacteria causing the infection. Given her expertise, Rountree had a set of standard phages that were well known and being used internationally to classify bacterial strains. She first tried these phages to see if there was a match. None of them worked. This was indeed a very different strain of staph. Rountree then modified the phage she was using, and her modified phage worked. But that introduced a new problem. She was now, apparently, the only person in the world with a phage that could identify this bacteria.

Rountree wrote to Robert Williams, a colleague at the UK labs in Colindale who was acting as the guardian of standard phages. She sent him the phage and the bacteria, and Williams tested both, also comparing her modified phage with the standard list. He confirmed that her phage was unique, and so was the strain.[12] But he was unconvinced that the outbreak of an infection in a pediatric ward in far-off Australia was anything of global importance. He gave Rountree's phage the rather dull designation of "80." (The number 80 was probably because it was the eightieth strain to be included in the growing set of phages at the lab in Colindale.)

Even as Williams did so, a similar outbreak was unfolding in Canada. Responding in much the same way as Rountree had, the Canadian scientists created their own phage to study the bacteria. And then, they too sent it over to Colindale. Once again, it failed to impress Williams, who called it 81. Subsequent research determined that both 80 and 81, while different phages, were pointing to the same infection.[13]

By 1956, the strain of resistant staph was not just in Canada and Australia. It appeared in New Zealand, England, and the United States. The global pandemic, the first of its kind due to resistance to a frontline antibiotic, made headlines in the newspapers and in magazines such as *Ladies' Home Journal*.[14]

The pandemic changed the way that hospitals, particularly infection wards, were run around the world. While Williams was initially unimpressed, and slow to respond, Isbister was the exact opposite. Once it was fully appreciated that resistance to penicillin was what caused the outbreak, she made sure that there would be an increased focus on how to prevent the rapid spread of resistance infection. In her ward, and subsequently throughout her country, she championed the idea of separating infected babies from their mothers, and created rules to minimize the length of time new mothers spent in the hospital postdelivery. The outbreak of resistant bacteria held immediate consequences for hospital infection control and hygiene. That could prevent the spread—but it wouldn't treat those who were already sick. What was needed was a new generation of antibiotics.

Panic set in as news of staph resistance went global. To the relief of many people, an antibiotic named methicillin became available in the UK in 1959.[15] It could not have arrived at a more opportune time. A semisynthetic drug with similarities to penicillin, methicillin seemed capable of doing what penicillin could no longer do. It worked exceptionally well for patients with penicillin-resistant bacteria.

As part of her job, Dr. Patricia Jevons of the PHL was culturing and studying staph infection strains sent to the lab from all over the United Kingdom. These numbered in the thousands. But in October 1960, she encountered something peculiar. Three samples that had come to her for testing from the same hospital in southeast England, were different from the rest of the batch. These three samples were showing signs of resistance not just to penicillin, and other first-generation antibiotics like tetracycline and streptomycin, but to methicillin.[16]

The first patient was in the hospital's nephrology ward. He had come in to have his kidney removed. The second patient was the nurse who cared for him. The third was someone who had visited the same hospital two weeks later. This third patient had not even been admitted but had been seen in the outpatient ward. After careful testing and retesting, Jevons came to a conclusion. Her published results opened with an ominous warning for those thoughtful enough to see it: "It is well known that patients with infected skin can be dangerous sources of infection in hospitals, and the finding of just such a patient infected with a methicillin-resistant strain in this instance adds an additional warning."[17]

Patricia Jevons's additional warning was initially minimized and overlooked. Many British scientists doubted her discovery of a new resistant bacteria. Their focus remained on the threat of penicillin-resistant staph and showed little concern over any methicillin resistance. Other reports of methicillin resistance from Eastern Europe and India also went largely ignored. But that was going to change.

About twenty miles south of the lab where Patricia Jevons worked is a large pediatric hospital, the first of its kind in England. Queen Mary's Hospital for Children had the unhappy distinction of being the most heavily bombarded hospital in all of London during World War II, and it was close to the Croydon Airport, which German airplanes targeted.[18] That its structure resembled an army barracks also didn't help. Evacuated during the years of constant threat of bombardment, the hospital reopened after the war and soon regained its strong reputation as a hospital known for treating and quarantining the infected.

In 1959, Queen Mary was one of only two hospitals in the United Kingdom using methicillin. And by 1961, just a few months after the publication of Patricia Jevons's warning, the doctors there noted the first case of methicillin-resistant staph in a baby.[19] Over the next few months it had spread to other wards, and even after adopting the new protocols of quarantine, the resistant strain remained difficult to contain, let alone treat. In 1962, the Queen Mary doctors reported their first fatality linked to this resistance. And soon the infection had spread to other parts of Europe.

Doctors in the United States were well aware of the reports coming in from across the Atlantic. But for some reason, America was spared. For nearly seven years after the first reports of this resistant strain came from England, there were no reports of methicillin resistance in the United States.

It wouldn't last. Initially, a leading Boston doctor by the name of Max Finland, working with his team, tested thousands of staph infections against all known therapies. There was evidence of resistance to penicillin. But methicillin seemed to work just fine. In early 1967, the vigilant doctor began another study, and this one stretched over a year. Only after the full study was complete did the team submit their paper for publication, which came out in 1968. The first line of the paper had an ominous warning. Twenty-two isolates from eighteen patients were found to be resistant to methicillin.[20] Methicillin-resistant *Staphylococcus aureus* (MRSA) had arrived in Boston, and soon it would be seen in hospitals across the United States.

CHAPTER 13

THE MAN IN THE BLUE MUSTANG

Max Finland "was so short you could only see his eyes when he was driving his shining blue Mustang," his former colleague Ron Arky remembered, chuckling.[1] Finland had been Arky's mentor and lifelong friend, and had also been the face of antibiotic resistance in the 1950s and '60s. His renown and accomplishments cast a far longer shadow than that, however. By the time Finland died in 1987, more than 60 percent of the heads of infectious disease departments in the United States had trained directly under him.

Though short in stature, Finland towered over the field, instrumental in shaping both the scientific discourse about and national policies governing antibiotics. Through his writing—over eight hundred scholarly publications and countless editorials in the *New England Journal of Medicine* (to which he never signed his name)—he shaped the field, and through his training he groomed an entire generation of infectious disease specialists. When the Infectious Disease Society was created in 1962, Finland was the obvious and automatic choice to be its first president.[2]

Finland was born in 1902 in a small village near Kiev, the capital of Ukraine. His great-grandfather had been the chief rabbi in the city of Krakow—and the family now lived in the Russian Pale. Created in 1791 by the Russian empress Catherine the Great, the Pale was intended to confine the country's Jews, who were not allowed to leave unless they converted to Russian Orthodox

Christianity. Poverty in the Pale was everywhere. Worse, conspiracy fears and anti-Semitism meant pogroms were recurring, especially during the latter part of the nineteenth century when it was believed by the Russian Imperial Court that Jews had been involved in the assassination of several prominent Russians, including the czar. The towns hit hardest by these waves of violent pogroms between 1903 and 1906 included the village where Finland was born.

Finland's father was not a man of great means, but he gathered what he could and led the family to safety. By the time they reached the United States, Finland was four. Their new home was a ghetto in the West End neighborhood of Boston, the city Finland would call home for the rest of his life. Finland entered Harvard College, graduating in 1922. He then attended Harvard Medical School, graduating in 1926. That he had been granted a scholarship spoke volumes; at the time, few Jews were so recognized. Finland's relationship with Harvard would last his entire life, and he more than returned Harvard's early faith in his talents: over the course of his career, through raised donations and his own frugal savings, he endowed multiple research professorships at the school.

Finland's first foray into infectious diseases started in the late 1920s during his residency at Boston City Hospital and the Thorndike Memorial Lab, which became his home for nearly half a century. Thorndike, created in 1923, was the first clinical research laboratory in the country and quickly became the epicenter for research and training. It was at Thorndike that Finland worked on what was then called the Captain of the Men of Death[3]—pneumonia—which accounted for nearly half the deaths at the hospital. It was also at Thorndike that Finland crossed paths with William Castle—a man who was almost his exact opposite in stature and personality. Finland was a dull lecturer; Castle was enthusiastic. Finland was short; Castle tall.

Thorndike was the perfect place to work for some of the most gifted physicians of the era. Finland had always been fascinated by infectious diseases, and starting in the 1930s, he turned his attention to pneumonia and sulfonamides—the blockbuster antibiotic of the 1930s that saved millions, including FDR's son, before it became impotent by the time of World War II.

Finland became a leader in the careful examination of penicillin's effects on the human body. At the height of World War II, there was a shortage of penicillin available for civilian use. Chester Keefer was appointed "penicillin Czar," responsible for rationing its use,[4] and he leaned heavily on Finland to provide him with advice on how the drug worked, when it worked and when it didn't, and how best to prioritize its use.

Finland was first and foremost a clinician, and was deeply troubled by the behavior of some of his fellow clinicians. He was among the most prominent and vocal advocates of the 1950s to argue for exercising caution when treating infections with antibiotics. Representing the United States at the World Health Organization alongside Finland was Dr. Selman Waksman. The two held starkly different outlooks. Waksman would argue that the world's natural resources would ensure an endless supply of antibiotics; Finland would sound the alarm, arguing that the world needed to be careful with these precious, life-saving drugs and avoid using them too often.[5]

In 1951, Finland wrote:

Are we in medicine, like our counterparts in industry, exhausting our most valuable resources at too rapid a rate? Are we, perhaps, depending too much on the ingenuity of our scientists and industrialists to keep constantly ahead of the decline? Only time will tell. In the meantime, is it not prudent to think more in terms of how these valuable antimicrobial agents can be used more effectively rather than most widely?"[6]

He did his best to rally the medical community, particularly surgeons, to avoid using antibiotics prophylactically. He had more than practitioners within his sights. Though Finland was often consulted by pharmaceutical companies, who rightly viewed the clinical trials he helped oversee at Boston City Hospital as both expert and efficient, he was never shy when it came to critiquing the drug industries for their aggressive marketing and less-than-transparent engagements with physicians.

In the mid-1950s, however, it was not just the drug companies that had adopted questionable practices. There was another institution deeply involved with antibiotics that was in the midst of a corruption scandal: the FDA.

Arthur Flemming, the cabinet secretary of Health, Education and Welfare in the Eisenhower administration, knew how Washington worked. His career in government had started nearly two decades before, in 1939, and he held broad bipartisan support. He was respected across the political aisle. But on June 6, 1960, he was in Senate hearings being raked over the coals.

The man sitting across from him was Senator Carey Estes Kefauver, a Democrat from Tennessee, and the senator did not care much about Arthur Flemming's distinguished career or his reputation among his peers. The issue at hand was a scandal in the antibiotics department of the FDA. And Kefauver was furious. How could Flemming and his team have failed so spectacularly?

The man at the center of the scandal was Henry Welch.[7] Welch had been at the FDA since 1939, starting around the same time Flemming commenced his own government service. In 1943, the US war effort had focused on producing truckloads of penicillin, and the army wanted to make sure that each batch met high quality standards. At the army's request, the FDA was tasked with ensuring that quality. A new division of penicillin control and immunology was formed within the FDA, and Welch rose through the ranks to become its head. As new drugs started coming to the market, including streptomycin (used for TB) and tetracycline (used for a variety of diseases ranging from cholera to skin infections), the mandate of the office broadened to include vouching for the quality of all antibiotics. In 1951, it was officially christened the Division of Antibiotics, and Welch retained leadership.

As the field grew, so did the need for making sure that relevant and valuable research reached the medical and scientific community. In 1950, Henry Klaunberg asked Welch to be the editor in chief of a new journal dedicated to cutting-edge antibiotic research. On the editorial board were the biggest names in the field: Selman Waksman, Alexander Fleming, and Howard Florey. Editing the journal was beyond the scope of Welch's governmental responsibilities, so he sought his superiors' approval before agreeing. He was also being asked to write a book on antibiotics, to be published by the Washington Institute of Medicine. The leadership at the FDA found neither to be problematic, readily agreed on both counts, and gave Welch permission. It was also understood that Welch would receive a modest honorarium for these tasks.

In 1952, the parent company of Welch's journal and the publisher of his book declared bankruptcy and sold the rights to the book back to Welch and to Dr. Félix Martí-Ibáñez, a recent Spanish immigrant who had become Welch's partner in his editorial responsibilities. The two men created two parent companies called MD Publications and Medical Encyclopedia Inc., which republished Welch's book with new branding and aggressive marketing. The *Journal of Antibiotics*, originally published by the Washington Institute of Medicine, also came under the control of Martí-Ibáñez, and while Welch was not an owner, his honorarium now depended greatly on how well the journal did.[8]

Welch and Martí-Ibáñez had a new business model: work closely, and collaboratively, with pharmaceutical companies. Deals were struck whereby they shared articles with drug companies ahead of their publication, and when those articles were favorable to the companies' interests, the companies would agree to buy a significant number of reprints. These companies also agreed to further support the journal through advertising. Conversely, articles that were critical of the industry were also shared ahead of publication, and many were subsequently rejected or their findings were toned down. Over time, additional journals under the same model were created and added to the business.

Welch's reputation and, more important, his position within the FDA as the top regulator of the industry, gave the initial journal both prestige and a semblance of objectivity. By 1958, Medical Encyclopedia Inc. was publishing more than ten volumes annually. The company's cozy relationships with the pharmaceutical companies increased. Welch began to solicit book ideas from them directly. In one instance, Pfizer told Welch that they might be interested in buying manuscripts and monographs that demonstrated the advantages of Terramycin, a drug Pfizer had developed. A book promptly appeared the next year on the topic, published by MD Publications.

MD Publications and the journal did phenomenally well between 1953 and 1959. Advertising revenues alone were upward of $300,000 during that period. Reprint sales were close to $700,000. And Welch's own earnings came close to $300,000. All the while, he retained his position at the FDA, where his annual salary was significantly lower than what he earned in profits through the journal.

By the mid-1950s, clinicians were raising concerns about the journal's pro–pharmaceutical company bias. There were murmurs about the lack of objectivity under the editorship of Welch. Some of the authors protested that the content of their papers was being changed to favor the pharmaceutical companies. The quality of the science and the unbiased role of the FDA were at stake, with a rising number of scientists awaiting the crisis coming to a head.

As part of its marketing campaign in mid-1950s, Pfizer had started using the term *third era of antibiotic therapy* to promote its new formulations. Welch, in 1956, used the same words as the head organizer of the Antibiotics Symposium. And when Welch delivered his speech at the gathering, the words not only made it into his opening remarks but into the prestigious publication *Antibiotics Annual*. It would turn out that Pfizer staff had authored the words and inserted them into Welch's speech, which helped ensure their publication in the annual.

The integrity of the FDA and the journal was in doubt, and so was the efficacy and ethics of medical practices. Confronting mounting evidence of antibiotic resistance—seen in penicillin and in other drugs—Welch started promoting the idea of fixed-dose combination, or combining multiple drugs into a single tablet. This was in line with what drug companies, particularly Pfizer, were promoting, and were referring to as "the third era."

Back in Boston, Max Finland was furious. He argued forcefully that there was absolutely no scientific or clinical evidence supporting the use of the fixed-dose combinations. He went so far as to call statements by Welch mere testimonials, and decidedly not scientific evidence.[9]

Concerns about Welch gained momentum—yet the stories about him remained largely confined to the community of infectious disease clinicians. In February 1959, John Lear, a journalist and editor of *Saturday Review*, broke the story for the general public. For his article, Lear interviewed Welch, who categorically denied any wrongdoing. He claimed there had been no conflict of interest. Welch told Lear that he had only received the honorarium that had been cleared by the FDA. He never discussed the sums of money he was receiving, or his interests in *MD Encyclopedia*.

Lear's article, and the questions it raised about Welch and the FDA, cast doubt about public confidence in the scientists and organizations tasked with vouching for the efficacy of antibiotics, and it created a furor. Letters from the public and from congressmen started flowing into Arthur Flemming's office. A formal investigation was launched by Flemming and progressed in fits and starts, due to Welch's declining health and to the FDA's disinclination to pursue the matter fully. Under ever increasing pressure, Arthur Flemming took action. In light of Welch's activities, Flemming created new guidelines for FDA employees.

The scandal paved the way for new policies in the FDA regarding ethical conduct and conflict of interest. While bacteria may be governed by Darwinian laws, the scientists and the pharmaceutical companies were driven by many other forces—including greed. The FDA learned the hard way that it had to act decisively to regulate not only the drugs but human behavior as well.

CHAPTER 14

HONEYMOON

For decades, soil scientists have been studying certain bacteria that are known to give some drinking water an earthy-musty taste and odor. This order of bacteria is known as actinomycetes—a Greek name meaning ray fungus.[1] They are named for their shape and uncanny resemblance to fungi. Actinomycetes can cause musty odors in soil as well.[2] They are a type of bacteria that has been known to outcompete other bacteria in the soil in their exploitation of its riches and resources. They help degrade organic matter and make compost, and they have an edge over other bacteria in decomposing insect and plant polymers. To soil scientists in the 1930s and '40s, this meant that actinomycetes were able to kill their competition.[3] Microbiologists have long wondered whether they could harness the arsenal of this class of bacteria to kill other disease-causing bacteria?

The man most associated with discovering the potential in actinomycetes—Selman Waksman—was buried in Crowell Cemetery in Woods Hole, Massachusetts. On Waksman's gravestone is an apt inscription from Isaiah 45:8: "The earth will open and bring forth salvation."[4] The words offer a scriptural insight. But they also reflect a scientific principle—one that was a touchstone to the work of the person buried there. All his life, Waksman had preached that soil would be the most prolific source of antibiotics. And he was proved right. Soil still continues to produce new molecules with unusual antibiotic potency. Waksman would pursue his belief with single-minded purpose, realizing life-saving discoveries, attaining extraordinary prominence—even a Nobel Prize—all the while denying credit to a scientist equally involved in those very discoveries.

Like Max Finland, Waksman was born in Ukraine to a Jewish family. He remembered growing up in "a bleak town, a mere dot in the boundless steppes."[5] Also like the Finlands, the Waksmans were driven out of the Russian Pale in part by the anti-Jewish pogroms of the late nineteenth and early twentieth centuries. Waksman arrived in the United States when he was twenty-two years of age, and in 1911, he began his studies at Rutgers University, from which he would graduate and spend nearly all of his career. From the beginning of his career, Waksman was interested in the therapeutic potential held by actinomycetes.

In the 1930s and early '40s, Waksman and his team at Rutgers made critical discoveries concerning the potency of various actinomycetes. A savvy salesman as well as a gifted scientist, Waksman was able to convince the pharmaceutical giant Merck to provide his lab with significant funding to finance the pursuit of new chemotherapeutic agents. The funds were critical to keeping the lab running, because the actual task of sifting through soil was laborious, frustrating, and often resulted in a dead end.[6]

The fortunes of the lab changed in the 1940s with the arrival of a new graduate student, Albert Schatz. Schatz was a workaholic, pushing himself often to the point of dangerous exhaustion. As World War II was raging, Schatz continued to work for Waksman, doing so even after he was drafted into the air force's medical detachment in late 1942. He took a pay cut and worked for what amounted to about a dollar an hour. Throughout it all, he remained motivated.[7]

Waksman had him testing soil samples for possible antibacterial agents that would target *Mycobacterium tuberculosis*, the same elusive discovery that Robert Koch had pursued a half century earlier. This was hazardous work, including the real threat of becoming infected with TB. Well aware of the dangers, and the fact that his lab was not safe, Waksman told Schatz to move the experiments to the basement to minimize the risk of any outbreak or exposure to himself. That basement became, quite literally, Schatz's home. He would work, sleep, and eat in the same lab.

His round-the-clock work paid off. On the afternoon of October 19, 1943, Schatz found a molecule, a hit, that could kill the TB pathogen. In fact, what Schatz had discovered was not a drug but a bacteria. Schatz named it *Streptomyces griseus*, reflecting its connection with the bacterium *Actinomyces griseus*. Both Schatz and Waksman knew they were shy of having discovered a new antibiotic, but both were certain they were on the cusp of doing so. More work was needed, and Schatz threw himself even more fully at the challenge. Finally, Schatz was able to distill what the duo believed was the essence of the bacteria—a molecule Selman named streptomycin.

Streptomycin was first tried on a pure culture of TB in Waksman's lab by Schatz, and then in vivo on lab animals. In November 1943, Waksman was visited by a team of scientists from the Mayo Clinic who got really excited about streptomycin. By February 1944, this became an active collaboration between Waksman's lab and Mayo—and Schatz watched from the sidelines as the work moved from Rutgers, New Jersey, to Rochester, Minnesota. The team of scientists at Mayo, including William Feldman and Corwin Hinshaw, first injected guinea pigs with TB and then gave them the drug. All four guinea pigs survived. Streptomycin was proving to be the next magic bullet. This was going to big, and Selman Waksman knew it.

Merck, initially uncertain about developing the drug, was converted to the cause by the ever persuasive Waksman. The results were everything he believed they could be, and over the next five years, with additional tests and trials confirming the potency of streptomycin, Waksman became a national and then an international hero. He had discovered the cure to TB, found in the backyard soil. Waksman was a genius to whom ordinary people could relate. The story was remarkably compelling. Here was a scientist, rising up from humble beginnings, an immigrant finding refuge in the land of opportunity, who unearthed the greatest of scientific treasures in the soil, one capable of saving humanity.

Schatz, on the other hand, was sidelined and forgotten. Now graduated, he fumed as Waksman received the world's acclaim, and all the while the older scientist failed even to mention the role Schatz had played in the discovery. When he complained to Waksman, the response shocked him. On January 28, 1949, Waksman wrote, "You know very well that you had nothing whatever to do with the discovery." In a letter sent just a month later, in February 1949, Waksman adopted an even harsher tone, scolding Schatz: "You must, therefore, be fully aware of the fact that your own share in the solution of the streptomycin problem was only a small one. You were one of many cogs in a great wheel in the study of antibiotics in this laboratory. There were a large number of graduate students and assistants who helped me in this work; they were my tools, my hands, if you please."[8] The dedicated hours spent at personal risk in the basement of Waksman's laboratory were minimized, if not completely erased.

Schatz's frustration was becoming something closer to fury, no doubt egged on by the tone of letters and the ever widening extent of Waksman's fame. Schatz filed a suit in federal court against his mentor, who up until that point was the sole recipient of the royalties for licensing streptomycin (Waksman's share was 20 percent, with 80 percent going to the Rutgers Foundation). When the case was settled, Waksman's royalties were reduced to 10 percent, with 3 percent going to Schatz and 7 percent split among everyone else who had worked in the lab during the discovery period. The compensation granted to Schatz by the court, however, did not translate into public acclaim for his role in the discovery. Waksman made sure of that.[9]

In 1952, Selman Waksman was the sole recipient of the Nobel Prize. He proudly accepted the biggest award in science and, adding insult to injury, in his speech he failed to mention Schatz even once. Waksman stood at the pinnacle of scientific renown, and Schatz, as a result of his public campaign against the national hero, was considered toxic. Despite his demonstrable skills and a track record of accomplishment, Schatz could not land a job at any top-rate institution. The scientific world snubbed him, but he kept campaigning to get due credit. That battle would continue well after Waksman had died.

The Philippine Islands reflect centuries of colonial occupation. Spanish domination began in 1566; American control began in 1898 and continued until 1942, when the islands came under Japanese occupation. The US forces retook control of the nation in 1945. The war was devastating to the islands, leaving the Philippines in desperate need of renewed investments in public health, education, and financial infrastructure.

Penicillin became the miracle drug of the war. With its success came a strong interest and investment from the pharmaceutical companies to discover and commercialize new antibiotics. The next decade was a honeymoon period for antibiotic discovery and commercialization, with scouts and researchers looking for new molecules all around the globe.

In 1948, Abelardo Aguilar started working in Iloilo, a central province of the Philippines. He was a local physician but had another job as well: medical representative of a pharmaceutical company called Eli Lilly, based in Indiana. Dr. Aguilar's task was to isolate promising soil samples and send them back to the company's laboratories. The company had recently expanded its international footprint and was looking for promising antibiotics. Like its competitors, Eli Lilly was literally turning over the soil of the world in its search.

In 1949, from soil in a cemetery in Molo, a district in Iloilo City, Aguilar extracted a drug from an actinomycete, similar to the one that Waksman and Schatz had worked on.[10] Excited about this discovery, Aguilar immediately took the extract to his company contact in Manila. His extract was one among many that were being sent from around the world to the Eli Lilly headquarters. But unlike the ones that were coming in from the other corners of the globe, there was something special about Aguilar's sample. The extract showed efficacy in killing bacteria that were proving resistant to penicillin.

Eli Lilly, upon confirming the results, immediately filed for a patent. The headquarters sent a letter to Aguilar thanking him for his remarkable discovery. To honor the site of the discovery, the new drug would be branded as Ilosone, though soon after it was broadly referred to as erythromycin. The world now had one more alternative to penicillin, and Eli Lilly had a blockbuster. In due course, erythromycin would become a major antibiotic used all around the world. By the mid-1950s, Aguilar felt that he was being denied his due credit. He started to press his claim, demanding royalties, and began a battle that continued for nearly forty years until his death. He was utterly unsuccessful.[11]

Eli Lilly's determination that Aguilar was not entitled to additional compensation was well understood by Reverend William M. Bouw, who died in 2006 in Toledo, Ohio, at the age of eighty-eight. A lifelong man of the cloth, Bouw had played a central role in the discovery of another blockbuster antibiotic that was owned and marketed by Eli Lilly.

Fifty-six years before his death, William Bouw had sailed across the Pacific with his wife and young daughter to reach the island of Borneo. Reverend Bouw had heard that there was a group of people called the Dayaks living deep in the jungle. He felt that it was his calling to convert the Dayaks from their ancient religion of Kaharingan to Christianity.[12]

Around this time, Eli Lilly had launched its own worldwide mission: to find new antibiotics from distant places, particularly in Southeast Asia. The success of the sample from the Philippines had further strengthened the company's belief that there were more treasures awaiting discovery, and they just needed individuals who were willing to locate and ship samples to them.

Bouw learned of Eli Lilly's mission through a messenger visiting his church. Believing the company's attempt to find even more life-saving drugs was compatible with his missionary work, Bouw decided to go deep into the jungle, not just to convert the Dayaks but to assist the drug company. By this time, he had friends among the Dayaks who were willing to act as his guides. With their help, he visited places where sunlight barely pierced the canopy of trees and where the soil was rich with organic matter. Bouw collected several soil samples and sent them back to Indianapolis. Thereafter, he forgot about the matter and returned to his missionary work.[13]

The samples from Borneo reached Eli Lilly headquarters, and a recent graduate from the Harvard chemistry department, Edmund Kornfeld, was assigned the task of analyzing them. The early results were remarkable. The soil sample contained a bacteria—*Streptomyces orientalis*—that could kill all staph infections, including those that were resistant to penicillin. Extracting the essence from the soil samples proved difficult. The process was long and required the meticulous use of dangerous chemicals. But Kornfeld persisted and extracted the active molecule. When dissolved, it produced a brownish liquid that the research team called Mississippi Mud. It was a name that belied the significance of their finding. The drug was dubbed vancomycin, from the word *vanquish*, and the FDA approved its use in 1958.[14] It was highly effective but failed to capture the market in the way the manufacturers had hoped for. It was edged out by another wonder drug—one that Max Finland had written about in 1968: methicillin.

Methicillin was the rock-star antibiotic, and soon after its arrival on the market it proved its worth. It saved the life of one of the most famous Hollywood actresses of the time: Elizabeth Taylor.[15] *Cleopatra* was the highest-grossing movie of 1963, but it almost never reached the theaters. And not because the film, originally budgeted to cost $2 million, came in at over $44 million. No, the reason was the near death of its leading female star, due to a particularly serious form of staph pneumonia. At one point during the course of Taylor's illness, her caregivers were told she had only one hour to live. Taylor's health was spiraling out of control. She would need a tracheotomy to help her breathe. The drug that saved her, and allowed the film to continue, was methicillin. That alone catapulted the drug to public awareness.

Methicillin was one of the first semisynthetic antibiotics—meaning that a naturally occurring antibiotic (in this case penicillin) was modified in the lab to create something that had a natural component as well as a part that was *not* present in nature, a synthetic part. It was not the first semisynthetic drug, but it was certainly the most successful product to come out of the UK's Beecham Pharmaceuticals in 1959.[16]

It was a product of bold experimentation in conjunction with an optimism about scientific possibilities held by pharmaceutical companies at the time. With substantial investments from drug companies, scientists working for them were coming up with new ideas and creating drugs using the building blocks of nature. Called the crown jewel by Beecham scientists, methicillin could treat penicillin-resistant staphylococci infections, a major problem in hospitals. The results of initial trials by Beecham were so impressive that it was fast-tracked to market. It took merely eighteen months to move from discovery to distribution, a process that would take nearly a decade today.

On March 12, 1961, the *New York Times* announced:

Behind the headlines last week about actress Elizabeth Taylor's
fight for life against staphylococcus pneumonia there is a dra-
matic story, known but to a few in the professional circles of
biochemists and microbiologists. It is a story in which the poten-
tial victim was not just one celebrated young woman but many
millions of men, women and children threatened by a deadly
new germ which has resisted penicillin and all other "wonder
drugs" of recent years.

The pipeline of new generations of drugs started to dry up by the
mid-1970s; the era of frequent discoveries of wonder drugs was
over. By the 1980s, the opportunities for pharmaceutical companies were in drugs that were not antibiotics. But there was one
major exception.

Bayer Pharmaceuticals' discovery of one of the most profitable
pharmaceuticals can be traced to the work of the Italian-German
scientist Johan Andersag. Andersag was working at Bayer in the
1930s when he discovered chloroquine. Animal tests suggested
that it was probably going to be too toxic for human use—and, as
a result, it was abandoned.[17]

It took a decade before scientists came back to the drug, tested
it comprehensively, and found its unique effectiveness in treating
malaria.[18] In the decade that passed before scientists returned to
testing of chloroquine, countless patients succumbed to the disease. In the 1940s, chloroquine was reintroduced by the Allied
forces, and the drug's ability to cure malaria was affirmed. This,
in turn, opened up researchers' interest in the capabilities of the
drug.

In 1962, scientists working on creating a more efficient synthesis of chloroquine discovered a by-product called nalidixic acid,
made up of a class of molecules called quinolones. The new compound had antibacterial properties and came to the market as a

therapy for urinary tract infections in 1967. Though effective, it would be another decade before another advance improved the drug that traced its roots to Andersag's discovery of the 1930s.

In 1979, the research arm of a Japanese company, Kyorin Sei-yaku Kabushiki Kaisha, applied for a patent for a derivative of drugs based on quinolones. The new molecule, for which the patent was granted, had a fluorine atom attached to its core, and was named norfloxacin.[19] It was highly potent, targeting methicillin-resistant bacteria, and was licensed to Merck in 1981. Other companies were trying to create their own versions of quinolones, and while other derivative molecules had been created, none were as strong as norfloxacin. In the wake of the addition of fluorine, a new class of antibiotics came to market. With a nod to quinine, quinolones, and fluorine, it was called fluoroquinolones.[20]

The efficacy of fluoroquinolones was substantially better than nalidixic acid. Spying a potentially lucrative opportunity, the large pharmaceutical companies started devoting vast sums of money to see what else could be done to fluoroquinolones. Just like their counterparts at competing companies, the scientists at Bayer were trying out all kinds of combinations and attachments to the molecule.

In 1983, Bayer scientists finally found something that was better than anything else yet developed. They published their data. The new molecule, Bay 09867, was nearly ten times more potent than any other fluoroquinolone.[21] More important, it was active against bacterial infection by a class of Gram-negative, rod-shaped bacteria called pseudomonas. Pseudomonas tend to live in water, soil, or damp areas and can cause infections in humans, including urinary tract infections.

The drug based on Bay 09867 was called ciprofloxacin, which quickly became commonly known as cipro. Introduced to the market just a year after norfloxacin, its sales soon dwarfed all competition, and by 2001, it was earning Bayer over $2.5 billion a year in global sales.[22]

But the most fundamental lesson went unlearned, the lesson

that had been spelled out clearly by Sir Alexander Fleming. The world received cipro as it had so many of the drugs that had come before. There was widespread enthusiasm—some even claimed that the era of resistance could well be over. They were, of course, wrong. Within a decade, cipro resistance was found in animals and humans in nearly every country in the world. Resistance to ciprofloxacin, and other molecules of the same class, was being widely reported in medical literature as early as 1990.[23]

The rapid and stunning advances of the scientific enterprise, which started in the 1940s and '50s and led to a steady stream of new and potent antibiotics that could treat a variety of resistant infections, was coming to an end. New drugs were difficult to make, and even the best ones did not last long before they became impotent. The scientific world needed to come up with other solutions. But before that could happen, another problem presented itself. Resistance was not just inherited from the previous generation of bacteria, it could also be acquired from completely different bacterial species. Bugs were becoming superbugs—they became resistant to many drugs simultaneously. The mechanisms controlling this development were unknown until a man named Joshua Lederberg came along.

CHAPTER 15

MATING BACTERIA

Joshua Lederberg came from a long line of rabbis and was expected to follow the family tradition, but he had different interests.[1] When he was ten years old, Joshua told his father that his heart was in science, not in Scripture. His father reassured him: All work done in the pursuit of truth, he said, is God's work.[2]

Over the course of his life, Joshua would make a series of discoveries that would transform the field of bacteriology. The first of these came early in his career, and it was also the one that would earn him the Nobel Prize. As a graduate student, Lederberg worked for his advisor, Edward Tatum, and the two men discovered a means by which bacteria can exchange material.

Through a series of simple and elegant experiments, Lederberg showed that bacterial cells perform something similar to sexual reproduction found in more sophisticated organisms. The phenomenon, called conjugation, occurs when two distinct types of bacteria come into contact, and the DNA of one, the donor, is transferred to the other, the recipient. The discovery of bacterial conjugation expanded bacteriology, in which it was understood that genetic information was no longer simply being passed down from parent cell to offspring cells, but was also being passed horizontally, from one bacteria to the other. This meant that bacteria could share their ability to resist an antibiotic not just with their offspring but with other bacteria that were previously susceptible to the antibiotic.[3]

Lederberg wasn't done transforming the world of bacterial genetics. With his graduate students and collaborators he made a series of discoveries in the early 1950s uncovering yet additional means by which bacteria exchange genetic information. This time, it was not through direct contact but through bacteriophage viruses. This means of transfer of genetic information was named transduction.[4]

Around the same time, in 1952, Lederberg coined the term *plasmid* to define DNA that could readily move from one cell to another. These DNA are not part of the nucleoid and not on the chromosome—and as a result can move vertically (from parent to daughter) as well as horizontally (from one bacterial cell to another).

Plasmids would fundamentally transform, and complicate, the phenomenon of resistance. This, however, was one discovery too many even for the extraordinarily productive Lederberg. The connection between plasmids and resistance would be made by researchers on the other side of the Pacific in the late 1950s.

Toshio Fukasawa cannot forget July 7, 1945.[5] The sound of the incendiary bombs dropped by US B-29s that hit his home when he was just sixteen are etched in his mind. In a matter of minutes, his family's house was burned to the ground. Also etched in his mind were the wounds his sister suffered, the burns that required her to be hospitalized for subsequent infection. Fukasawa recalls that it was penicillin, produced in the basement of the Chiba Army Hospital, that saved her life.

Ever since then, Fukasawa's relationship with antibiotics was both personal and professional. Reflecting on the devastation in his beloved Japan, Fukasawa initially wanted to become a nuclear scientist. He wanted to figure out why and how so many people in Hiroshima had died from a single bomb. But when he failed the entrance exam to become a nuclear physicist, he reconsidered his interests. Medicine was next on his list, and this time he passed the necessary exams. Soon after finishing medical school, he took on an unsalaried intern position with a quiet, soft-spoken man named Tsutumo Watanabe.

Infectious diseases were a major challenge in postwar Japan, with dysentery being a repeat concern due to the war's disruption of the country's infrastructure. But what made its repeat outbreak especially concerning was sulfonamide resistance. As of 1957, Japanese scientists knew that the strains causing bacillary dysentery were resistant not just to sulfonamide but to a broad swath of first-line antibiotics. For example, a Japanese man who had lived in Hong Kong then returned to Japan in 1955 showed signs of dysentery. More alarming, he did not respond to the standard treatment of antibiotics.

Familiar with the rising extent of drug resistance, his doctors decided to culture bacteria from the patient. To their surprise, they found that the bacterial strain was resistant to four different drugs: streptomycin, tetracycline, chloramphenicol, and sulfonamide. Prior to 1955 no such strains exhibiting resistance to four different antibiotics had been reported in Japan. But over the next few years such cases increased, and between 1957 and 1958 such resistant cases would quadruple in Tokyo alone.

Despite Lederberg's work on bacterial conjugation, few scientists were sufficiently familiar with it, and consequently very few suspected that resistance could transfer from one type of bacteria to another. This was about to change. Fukasawa dates the change precisely: the annual meeting of Japanese Society for Bacteriology in 1959. A fellow scientist named Sadao Kimura gave a presentation of recent research that showed results that were astonishing. Indeed, many attendees were incredulous.

Kimura's experiment was simple. He mixed two different types of bacteria. One was a strain of *Shigella* bacteria, which had shown resistance to tetracycline, chloramphenicol, streptomycin, and sulfonamide. The other was a strain of *E. coli*, which was sensitive to all these drugs. Kimura left the mixed culture overnight. By morning, the *E. coli* strain was resistant to the same drugs as the *Shigella* strain.

Prior to these experiments, most of the world's scientists believed that bacteria become resistant by random mutations, and then the mutant strains, under the assault of antibiotics, survived and passed the mutation down to their offspring. Strains that are not resistant die out; strains that are resistant survive. Eventually, only resistant strains are left, but this process was presumed to be slow and random. The idea that resistance could transfer between two bacterial species over the course of a single night was beyond belief.

Kimura's initial assumption was that some kind of transduction, or a process carried out through bacteriophages (the process, that is, that had been discovered by Lederberg), was going on. To test this hypothesis, Kimura did another experiment that was devised to ensure that if the *E. coli* picked up the *Shigella*'s resistance, it would be able to do so only due to the viruses, not the strain. The result was negative. Phages were not involved. Something more direct was going on between the two distinct bacteria.

Kimura's fascination was piqued, as was his concern. He set out to determine whether the phenomenon between these two strains was a one-time observation or was present in other cases. He sought out other strains from other employees at his university. Some of them had resistant *E. coli*, and some of them had *E. coli* that were sensitive to drugs. Kimura mixed the two strains. When he checked the results the next day, he found that all strains had become resistant.

Kimura's results challenged the existing dogma. But he offered no clear explanation of what was going on. Still, Fukasawa was stunned, as were others in the audience. The results meant two things. First, strains of bacteria could infect other strains. Second, and more troubling, was the possibility that bacteria could become resistant to drugs that hadn't even been used yet to combat them. Put bluntly, the bacteria had a mechanism to stay one step ahead of the antibiotics.

Fukasawa went to his boss, Tsutomu Watanabe. Watanabe was intrigued. First, the scientists decided to repeat Kimura's experiments. They went to Japan's National Institute of Infectious Diseases, which had a large collection of strains of multi-drug-resistant *Shigella*. They took those samples and combined them with their own, nonresistant strains of *E. coli* and *Salmonella typhimurium*. Their results replicated what Kimura had reported. Resistance was indeed transferring from one strain to another.

Over the course of the following year, Watanabe and Fukasawa undertook a series of experiments to figure out precisely how this was happening. Was this actually a genetic transfer, through cell-to-cell conjugation, as seen by Lederberg over a decade earlier? The answer was unequivocal. The culprit was plasmids, the self-replicating DNA molecule that could pass from one bacterial cell to another. They named the plasmids the R (for resistance) factor.

Published only in Japanese, and read almost exclusively in Japan, their findings did not immediately reach a Western audience. To rectify that, during the 1960s, Watanabe wrote a series of articles that showed for the first time, to Western audiences, how R factors were responsible for multi-drug resistance.[6]

Their findings were a jolt to the global research community. Resistance was no longer just inherited, or acquired through random mutations. To become resistant, a particular bacteria didn't even have to come into contact with the antibiotic. Bacteria could become resistant just by coming into contact with another bacteria that was already resistant to one, or even multiple, antibiotics. It was no longer Darwinian evolution alone that made the bacteria resistant. It was more complicated than that. A new front had opened up in the battle between pathogens and potent drugs. Tackling resistance was no longer a simple pursuit of making new drugs. It required a deeper understanding of another field of biology: genetics.

CHAPTER 16

S IS FOR SOVIET

The year was 1948, the month September. A new Cold War, in the shadows of World War II, was commencing. The world would soon be cleaved in two, an area of Western influence and an area of Soviet influence. At this time, Soviet science was in tumult. The science of genetics was being declared not just a bourgeois idea but, in fact, a threat to Soviet society. The man driving the accusations against genetics was Trofim Lysenko.[1]

Though not formally trained in genetics, or in any science, for that matter, Lysenko managed to give a uniquely Soviet spin to genetics. His focus wasn't on infections or the human body, but on agriculture, and his primary claim, which was deemed ideologically palatable, was that the highest, richest crop yield would result from exposure to the right conditions, even in the toughest of environments.

That Lysenko supported the Stalinist policy of forced collectivization of farms, which was devastating Soviet agriculture, substantially helped secure his stature within Russia. And that Lysenko was ruthless, ambitious, and ready to charge anyone who supported Western scientists as supporting imperialist propaganda made him a person to fear.

Marxist ideology was at odds with the primary findings of genetics, at least as the subject was studied and then understood in the West. Its emphasis that beneficial traits would pass down from one generation to the next could be understood to imply that the proletariat would always remain genetically subservient to the bourgeois, because the bourgeois had inherited more desirable genes. Less important than the accuracy of that belief as a matter of science were its political implications. And to the satisfaction of Soviet Marxists, Lysenko refuted it, stating that such inheritance was not how nature worked. Instead, he declared that nature could be trained under the right conditions. The political implications were equally clear. Lysenko and the proponents of his ideas argued that just as plants, upon exposure to the right conditions, could produce optimal outcomes, no matter how hostile and unsuitable their previous conditions had been, the same was true when people were exposed to the right ideology.

Lysenko's radical ideas were attractive to Stalin and his inner circle, and they cared little that he had no formal scientific training. Indeed, that these ideas came from a commoner suggested that they were ideologically perfect and completely sound. Further cementing Lysenko's stature among party hardlines, scientific scrutiny was less important than Western opposition to his ideas.

The more the West claimed Lysenko a pseudoscientist, the more resolved the Soviet Union became in its support of him. By the late 1930s, Lysenko had Stalin's full backing, and he used it to eliminate any serious scientific questioning of his findings, which included discrediting Nikolai Vavilov, a renowned biologist, the father of Soviet genetics, and at one time Lysenko's mentor. Vavilov was highly regarded among the international community, but his persistent opposition to Lysenko became intolerable to the Soviet state. He was arrested in 1940 and, in a cruel irony (he was the person most credited with revolutionizing Soviet agriculture), he died in prison of starvation in 1943.

Arguably, the height of Lysenko's influence was reached in September 1948 when he was presiding over a special session of the presidium of the Academy of Medical Sciences.[2] A resolution was brought forward that anyone who disagreed with Lysenko-ism should be expelled from the academy. The resolution had an explicit target, a man who was not even in attendance. The head of the Antibiotics Laboratory in Moscow and a member of the Soviet Academy of Sciences, Gregory Frantsevich Gause was already working under a cloud.[3] He had been accused of being a British spy by the Soviet newspaper *Pravda*. Many thousands had been arrested and executed for less.

Gause was well known in Soviet science by the 1940s. He was highly regarded for his work in ecology and evolutionary biology. In the mid-1930s, he had advanced a theory about the competition of species, what he called the struggle for existence. Gause had proposed, correctly, that two species competing for the same limited resources could not maintain their population levels forever. At some point, the one with an advantage, no matter how tiny, will win over the other.

By the late 1930s, and near the outbreak of World War II, Gause had turned his attention to antibiotics. In 1942, working alongside his wife, Maria Brazhnikova, Gause discovered a new antibiotic. Produced from the bacterium *Brevibacillus brevis*, the new drug proved to be remarkably effective in killing *Staphylococcus aureus*. Gause named his discovery Gramicidin S—the S added as a mark of respect and recognition that it was a product of the Soviet. In 1946, he received the highest civil award, the Stalin Prize.[4]

Along with penicillin, Gramicidin was a major resource for the Soviet Union during the war. In desperate need for effective ways to treat infection, its production was fast-tracked and used throughout Russian hospitals. It was also sent, through the Red Cross, to the United Kingdom in 1944. The reason for this shipment was the exigencies of war. The Soviet Union needed to have the structure of the drug determined so that it could be produced in higher quantities and in a purified form. However, four years later, this shipment to England was used as proof of Gause's having colluded with the imperialist powers and the enemies of the state.

On the same day the Soviet academy's resolution to expel him passed, Gause received a call. The caller did not identify himself, and Gause did not bother asking who he was. The message was more important than the messenger. It was simple enough: "Keep going to your office; you are safe."[5] Gause showed up the next day at his lab and continued to do so for the next four decades. While a shadow of ideological suspicion always hung over him, Gramicidin S was too important for Soviet state security, and, consequently, the scientist was allowed to continue his work without interruption.

Though he steadfastly refused to become a member of the Communist Party, which meant he was overlooked for the top science positions in the Soviet Union, Gause's work on antibiotics went on throughout the 1950s, '60s, and '70s. His research continued to gain international attention, while Lysenko's ideas faded into ever greater obscurity. And at the time of Gause's death on May 2, 1986, Gramicidin S was among the most widely produced antibiotics in the USSR.

Just as greed continued to influence science in the West, the contest between politics and science was a stark reality in the Soviet Union. Gause's discovery, its importance, and his consequent fame provided him with some protection from the politics of the day. Soviet suspicion of scientists and state interference in scientific research would remain a hallmark of not just the Soviet Union but of scientific and medical research undertaken in the Soviet satellite states in Eastern Europe. Nowhere was this tension between science and politics more noticeable than in the German Democratic Republic—better known as East Germany. Sometimes, however, in the larger struggle between bacteria and humankind, the most politically restrictive societies can have unintended silver linings.

In 1978, a young officer from the dreaded East German Ministry of State Security, or the Stasi, came to Wolfgang Witte's lab to interview him.[6] Though Witte had been alerted that he should expect the visit, he was necessarily on his guard. The visitor had questions about Witte's recent visit to Mongolia—a Soviet satellite state at the time.

Specifically, the officer wanted to know Witte's opinions regarding the vaccinations against *Staphylococcus aureus* infections that the Soviets were administering in Mongolia. The officer reminded Witte that the efforts to control infections in Mongolian hospitals were based on recommendations coming out of the Soviet Union. Then the officer got to the point: Why was Witte criticizing these programs? The official slogan in East Germany at that time was: "To learn from the Soviet Union means to learn to be victorious." Did Witte not know that?

Witte did know that—but he also knew that the data on the vaccine's efficacy was bogus, and that the Soviet claims that the vaccine produced few if any side effects were fabricated. Witte and his own boss had been vaccinated using the "superior" Soviet vaccine and knew firsthand that the side effects—high fever and persistent chills—were awful.

Witte also knew that the Soviets performed unethical clinical trials with its vaccines in orphanages and in prisons, where resistance to TB drugs was particularly high. And he was aware that during a confrontation with a Stasi agent he should keep his mouth shut.

Witte was already regretting a conversation he had had some time before with a fellow passenger on a long flight back to East Germany from Mongolia. Witte had been relaxed, and their polite conversation had turned to a discussion about work. He had shared his thoughts about the Soviet Union's vaccine campaign. It was now clear that the passenger had been the Stasi's source of information.

The interview with the Stasi agent was short, and Witte kept his cool. He knew that every word was being recorded. He could even see the tape recorder that was woven into the lining of the officer's jacket. Thinking on his feet, and taking a chance, Witte told the officer that, as a good socialist, he must be truthful; if citizens are not critical of ineffective programs, then they are dishonest to the true spirit of Marxism. It was a gamble—speaking up typically didn't end well in East Germany. But this time it worked.

The Stasi officer left, and no one followed up with Witte. Whether it was due to Witte's smart response, the accuracy of his observation, or his stature is unclear. What is clear: Witte by this time was a well-known scientist working on the problem of antibiotic resistance. And what he was uncovering was of far-reaching significance.

Born in 1945, Witte grew up fatherless in a village in the Harz Mountains and attended the local school for the gifted, known as the gymnasium. He went on to study biology at the University of Halle. He focused on scientific disciplines but also took compulsory courses in Marxism-Leninism. According to Witte, there were two groups of teachers: those who were educated and knew their technical material, and those who would regurgitate Marxist ideas without understanding them. Discussions with the latter group were dangerous, due to the presence of informers and spies among them. These informers were always keen to report any dissent or criticism of policies as anti-state activities.

By the mid-1960s, the study of genetics—suppressed as a bourgeois topic in the 1940s—was experiencing a revival in the Eastern bloc. Many scholars now considered it to be a robust area of study within the field of biology. And by 1969, when Witte was asked to stay on as an assistant professor in the Department of Genetics at the University of Halle, microbial genetics was taking off in East Germany.

But the Communist Party's influence and control was even stronger. The 1968 crushing of the Prague Spring had soured the East German government of Walter Ulbricht and greatly increased its anxiety over student movements. As the student protests in West Germany roiled campuses across the country in

1968, the authorities in East Germany cracked down. The relatively free atmosphere that Witte had experienced as a student at Halle University had changed. Restrictions replaced intellectual freedom. And whenever possible, according to the mandates from the Communist leadership, scientific lectures should reflect Marxist philosophy.

Witte imagined that he would not last long in this new climate at the university. To be promoted and enjoy a career as a professor, you had to be a member of the party, or at least have come from a working-class family. Witte could claim neither. He wasn't interested in joining the party and actively pushed back when the head of the local chapter urged him to join.

Witte knew that his opportunities at Halle were shrinking and that he had to move. As luck would have it, a student of Witte's was the son of Helmut Rische, who happened to head the Institute for Experimental Epidemiology (IEE or Institut für Experimentelle Epidemiologie). A job at the institute opened up in 1973. It was in Wernigerode, a town not far from where Witte grew up. He applied for and got the position.

The institute, which was part of East Germany's national system of public health, originally focused only on public hygiene. Over the years, it had become the national center for epidemiology and the analysis of antibiotic resistance. While East Germany could boast advantages, especially economic, compared to other Soviet bloc countries, it was well behind its western twin. This was true in many respects, including science. While the pharmaceutical industry was thriving in West Germany, it was weak in the east. Compared to dozens in West Germany, only two pharmaceutical manufacturing facilities in East Germany—one in Jena and one in Dresden—produced a few antibiotics. Drugs that had already been introduced in West Germany, like trimethoprim, would take years to reach East Germany. But most drugs that made it into East Germany came from elsewhere. Erythromycin came from Poland, oxacillin from Russia, gentamicin from Bulgaria. In the late 1980s, cefotiam came from Japan and ciprofloxacin arrived from West Germany. Indeed, the disparities between the east and west were marked in terms of availability of essential and potent antibiotics. For example, glycopeptides, which were easily available in England, the United States, and West Germany, were rarely available in East Germany, which in turn affected how patients were treated.

The health authorities in East Germany were aware of their shortages of drugs and created a strict prescription regime as a result. Antibiotics fell into three classes. Class A drugs could be prescribed easily at any time and were readily available with a prescription from the local pharmacy. Class B drugs required a senior physician to sign off on the prescription, and the prescribing doctor was required to fill out a significant amount of paperwork. Class C drugs required a prescription from the senior leadership within a hospital, and the drugs were almost always imported. The result was that patients who needed these second-line antibiotics often didn't receive them (unless, of course, the patient was a member of the Communist leadership).

Just as he knew that the shortage of antibiotics was a serious issue, Witte also knew that antibiotic resistance was a growing problem. Researchers had known since the early 1950s of *Staphylococcus aureus*'s increasing resistance to antibiotics.[7] More recently, infections had become increasingly common at the state's maternity homes. Concerned about the declining birth rate in the years following the end of the war, East Germany had rolled out incentives for citizens to have more children. The incentives worked, and pregnancy rates increased. But the government had insufficient funds to improve the hygiene conditions of the increasingly overcrowded maternity homes. This lack of funding led to high incidences of mastitis, an infection that creates inflammation in breast tissue. With crumbling infrastructure due to limited funds, lack of hygiene and overcrowding led to the rapid spread of infection. Initial results showed that it was due to a multi-drug-resistant strain of *Staphylococcus aureus*.

The IEE was in charge of classifying the strains and researching the outbreaks. Witte was in charge of studying the causes of the outbreaks, and he spent much of his time looking at clinical samples from patients in various hospitals across East Germany. His lab became the country's reference center for staphylococci. Central to his work was tracing the source of the antibiotic-resistant *Staphylococcus aureus* strains that were cropping up in multiple hospitals. His team also investigated the troubling evidence that the strains were jumping from the hospitals to the local community.

Witte's group soon established a network among diagnostic laboratories across the county so that they could maintain surveillance of antibiotic resistance of bacterial pathogens. There was, however, a recurrent challenge. Due to limited financial resources, sufficient diagnostic devices that would measure the susceptibility of a bacteria to a particular antibiotic, produced by well-established international providers, could not be imported, and they had to be manufactured locally in East Germany. These devices were not always dependable, and the results were inconsistent.

There was yet another challenge for Witte and his team. They knew that the best vetting of their findings would occur if they could publish them in Western scientific journals. But before Witte could turn to a Western journal, he had to demonstrate that a similar paper had already been published in a journal in East Germany or, ideally, in the USSR. Politics demanded that new findings be published first in Soviet journals. Despite these burdens, IEE kept pushing forward.

Witte's interest in resistant *Staphylococcus aureus* in both humans and nonhuman animals placed him at the forefront of East Germany's research efforts, and it was the reason why, in 1978, he was given the chance to go to the Mongolian People's Republic, a remote Communist satellite of the Soviet Union. It was a poor country with minimal resources. Access to drugs in Mongolia was even more limited than it was in other Soviet satellites.

Witte crisscrossed the country collecting samples to try to determine the extent of drug resistance in Mongolia. To obtain them he had to receive permission from the local security apparatus, a process that was bureaucratic, extensive, and often opaque. But Witte had a letter of permission from the chief of the Mongolian police. With the letter tucked in his old leather jacket, Witte headed back to East Germany. As he was waiting to board his flight home, guards asked Witte for authorization for the samples. Witte produced the letter. The guard looked at it, folded it, and told Witte to wait. Soon, two guards with Kalashnikovs arrived and told Witte to accompany them.

Witte was certain that this would end poorly. But state security was so impressed to see a letter by its chief of police that they treated Witte as if he were an incredibly influential person. He was given reserved seating, for dignitaries only, while he awaited his plane. Witte's initial concerns dissipated, and he began to relax when he was on the flight. That was when he opened up to his fellow passenger, the informant. The casual conversation with the informant, sitting next to Witte, eventually led the Stasi to visit Witte's lab.

Witte found reduced resistance to antibiotics in his samples from Mongolia; the poverty of the country had prevented the availability or wide use of antibiotics, and, as a result, the bacteria there were less likely to develop resistance. Mongolia was also remote and had few interactions with the rest of the world—and resistant infections were unlikely to arrive due to its isolation.

While East Germany was poorer than its western neighbor, it was one of the richest countries in the Communist bloc,[8] far wealthier than Mongolia. Among the few agricultural industries that thrived in East Germany, pork was at the top. It exported pork not just to the Communist bloc but also to countries on the other side of the Iron Curtain, including West Germany, France, and England, securing hard currency for East Germany.

For Witte, this phenomenon posed a challenge. East Germany's pork industry exclusively used a feed that contained the antibiotic oxytetracycline, which the industry added to promote faster growth in the animals—more meat in a shorter length of time. This antibiotic was used for treating patients as well. Witte was certain that this broad use of the antibiotic was leading to the widespread resistance to the drug in humans and in animals. Witte and his team's theory was roundly challenged by the industry, but they made their concerns known. Eventually, Witte and the IEE succeeded. The IEE was able to convince the government to ensure that the antibiotic used as a growth promoter in pigs was not also used to treat humans. Thereafter, pigs received oxytetracycline, an antibiotic that was not used in humans and exhibited no cross-resistance to antibiotics either.

The privations of East Germany forced physicians to use the available antibiotics in a prudent and rational way and to employ infection control measures in order to prevent dissemination of resistance. The IEE also helped form a single government body that decided how antibiotics would be used in human and veterinary medicine as well as in agriculture.

Not much went right in East Germany. The regime was brutal, repressive, and authoritarian. But some policies in public health did pay off. Because of their limited resources, the East Germans created and implemented a policy of "One Health"—the single framework of looking at animal, human, and environmental health. This One Health idea would become popular in the West three decades after the East Germans came up with it. The dividends of some of these policies were clear. At the time of German reunification, resistance rates in East Germany were significantly lower than that of its western and significantly wealthier twin.

CHAPTER 17

THE NAVY BOYS

On the other side of the Cold War divide, and almost literally on the other side of the world, a young doctor, Lieutenant King K. Holmes, was also confronting the challenge of mounting cases of drug-resistant bacterial infections. Through his research, Holmes realized that discovering new ways to prevent the spread of infections was just as crucial as discovering new lines of antibiotics.

Holmes was a recent recruit to the navy. A graduate of Harvard, he went on to secure his MD at Weill Cornell, followed by a prestigious internship at Vanderbilt University.[1] While he was at Vanderbilt, Captain Herbert Stoeckline, a navy recruiter, called Holmes to let him know that he was soon to be drafted for the war in Vietnam. Holmes was slotted to be sent to the Mojave Desert to be part of the Public Health Service Commissioned Corps. But Holmes was not interested in the public health corps, so he made a counteroffer. He told Stoeckline that he was willing to work for three years, instead of the two required by the government, if he could be stationed in Hawaii or the Philippines, where he could work on infectious diseases. Stoeckline agreed.

So during the war, Holmes was stationed on the USS *Enterprise*, a naval vessel with an illustrious past. The name has been reserved for special ships over the course of US history—eight American naval vessels have been given this name, beginning

with a British ship that the Americans captured in 1775. At the
height of the Vietnam War, the name had been given to one of
the prides of the US fleet, the first-ever nuclear-powered aircraft
carrier deployed in the Pacific. Holmes was assigned to the preventive medicine unit on the USS *Enterprise*, where the doctors encountered the problem of recurring infections among the sailors.

Holmes was required to spend part of his time in Pearl Harbor
and part of it on the ship. When in Pearl Harbor, he located an
old Harvard instructor, Claire Folsome, who was running a microbiology lab at the University of Hawaii. When King asked to
join the lab, Dr. Folsome gladly agreed. And while there, King
started learning epidemiology from researchers at the University
of Hawaii.

Before long, the USS *Enterprise* was docked at Subic Bay in the
Philippines. Subic Bay is about sixty miles northwest of Manila
and is about the size of Singapore. By the time Holmes arrived,
the war in Vietnam was raging and the bay had been transformed.
It was not just a naval station built by the US military; it was a
world unto itself.[2] One of the largest stations for logistics ever
built, it was the hub of wartime transportation and military coor-
dination. Which meant it was teeming with military personnel.
Over forty ships at a time could dock there, and in 1967 alone,
more than 4 million sailors visited the bay on their way to the
war, or on their way back home.[3]

Such a large number of servicemen in the region was a boon
for local trade and commerce. The navy servicemen who were off
duty would go into the towns and visit the bars, nightclubs, and
brothels, sometimes for entire weekends. Once back on the ship,
however, many started complaining of a urethral discharge that
the naval doctors soon diagnosed as gonorrhea. By the time King
reached the *Enterprise*, the number of infected servicemen was
escalating, and people were concerned.

Operating under the assumption that all of the cases were indeed cases of gonorrhea, the doctors followed the standard treatment protocol of the time and prescribed penicillin to the afflicted servicemen. But when King took up his post, nearly half of the servicemen weren't responding to the drug.

Holmes found the failure rate quite troubling. He collected samples of urethral discharge and took them back to his old lab in Hawaii. As he studied the culture of the discharge to understand why the standard penicillin was ineffective, he was surprised to find that about half of the discharge was not gonorrhea at all—it was, in fact, nongonococcal urethritis, an infection caused by pathogens other than gonorrhea. The half that *was* gonorrhea, however, exhibited signs of serious drug resistance. The treatment failures that the doctors witnessed at Subic Bay were due in part to misdiagnosis and in part to drug resistance.[4]

Holmes figured out the first part of the problem easily. Some of the men presented with symptoms similar to gonorrhea but were actually suffering from an infection already known to be insensitive to penicillin. These patients simply needed a different drug.

The bigger question was why those servicemen who had contracted gonorrhea were not responding to penicillin. Holmes had to be more than a preventive medic, more even than a microbiologist. He would need to become a field epidemiologist—in other words, a disease detective. What Holmes would do next would go far beyond US Navy guidelines in Subic Bay—it would change the guidelines for the treatment of drug-resistant gonorrhea throughout the world.

Holmes started his field investigation in one of the major settlements on Subic Bay. Olongapo was home to nearly five thousand sex workers, and every day about 250 of them would appear at the one clinic in town for a monthly health checkup. The patients would line up to see the one female doctor running the clinic.

Holmes knew that if he hoped to understand the reasons why the disease was spreading among the navy men, he had to start at this clinic, where the sex workers were being screened.

The local doctor used a single speculum, a standard tool for vaginal examinations. For efficiency's sake, she kept a large bucket of water nearby to rinse off the speculum after she had completed an exam. She'd then continue in this way, seeing patients one by one, rinsing the speculum in the same bucket of water. As he observed the doctor's clinic, King had found his first clue. The doctor's examination practice, and especially that bucket of water, was a major source of the infection that was spreading among the sex workers and the sailors.

This explained the spread of the infection but not the bacteria's resistance. Holmes turned to the medicines that the doctor was giving to the women. The standard treatment for those who showed signs of gonorrhea was benzathine penicillin, rather than procaine penicillin. The difference between the two forms of the drug is that benzathine penicillin releases a low dose and stays in the patient's system for a long time, while procaine penicillin releases a high dose and is out of the patient's system earlier. Benzathine penicillin was ideal for treating syphilis because of the lower likelihood of resistance. Treating the disease needed a long-acting drug and not one that would leave the system quickly. On the other hand, gonorrhea, due to increasing drug resistance, needed a fast-acting drug that would not stay in the system for long and create an opportunity for selection of resistant bacteria. Holmes knew how both drugs functioned in vivo and was troubled by the fact that gonorrhea patients were given benzathine penicillin as opposed to procaine penicillin.

Holmes had found his second clue. He saw that the sex workers were prescribed the wrong type of penicillin, and that snowballed into a major spread of Neisseria gonorrhoeae in Subic Bay. Drug resistance meant that the bacteria would survive a low dose of the drug being prescribed.[5] Holmes had to check for one more thing before he would be sure of the cause of resistance. He went to community pharmacies and asked if sex workers were coming in requesting antibiotics on their own (such as benzathine penicillin), without a prescription. His hunch was right. And the pharmacies were readily providing the sex workers with benzathine medicine. Not only was the infection being spread, but the sex workers were prescribed the wrong medicine, which led to resistance. The problem was exacerbated by the practice of self-medication and overuse of the same wrong medicine.

Now that Holmes had solved the puzzle, it was time to head to the *Enterprise* and report on his findings. He promptly made three recommendations. First, he made the navy buy nearly 250 specula, one for each patient the doctor saw over the course of each day. He also had the navy buy an autoclave, a machine that would sanitize all the instruments overnight so that they would be ready for use the next day. His third recommendation was to institute a method of Gram-staining on the ship so the doctors would know which bacteria was responsible for the infection. This would quickly ensure the proper diagnosis of the servicemen and, thus, better treatment.

If the patient had nongonococcal urethritis, then he would get tetracycline. If he *had* gonorrhea, then he would get procaine penicillin combined with probenecid, or tetracycline combined with probenecid. Holmes recommended this addition because penicillin, while effective, is often excreted quickly, and the presence of probenecid ensured that the high dose of the drug stayed in the system long enough to treat the infection completely. Once the new system was instituted, the sailors started to get better. The number of patients with drug-resistant gonorrhea plummeted, and cross infection within the examination room sharply declined.[6]

War and conflict have always been harbingers of injury and infection, and the desire to save the lives of soldiers engaged at the front lines has often been a catalyst for medical advances. Indeed, investigations like Holmes's proved essential in the contest between humankind and bacteria. By 1967, he felt a sense of accomplishment about his discovery. His findings led to infection-screening protocols and specific combinations of drug prescriptions that are used to this day. And yet drug resistance remained a fundamental challenge as bacteria continued to evolve.

Holmes had developed a special bond with the Philippines and its people during his time there in the 1960s. While the Vietnam War was over when he decided to go back to the island nation in the 1990s, its aftereffects continued, in some obvious ways, and in other not-so-obvious ways. Holmes and his colleagues went back to further investigate antimicrobial drug resistance among sex workers. He wanted to see whether the patterns of disease had spread, and whether access to health care among the sex workers of the 1990s was different compared to that of the 1960s. The team found that in a two-year period, from 1994 to 1996, resistance against one of the most potent antibiotics of the time, ciprofloxacin, for both Gram-positive and Gram-negative infections, had increased from 9 percent to 49 percent among Philippine sex workers.[7] The drug that had been touted by Bayer as a solution to resistance was now proving less and less effective.

CHAPTER 18

FROM ANIMALS TO HUMANS

The Paraná is the second largest river in South America, second only to the Amazon. It is formed by the confluence of the Paranaíba and Rio Grande in Brazil, and as it flows southwest, it creates the natural boundary between Brazil and Paraguay, and between Argentina and Paraguay. As it zigzags into central Argentina, it goes from the city of Rosário to the nation's capital, Buenos Aires.

For over half a century, the citizens and industries of Rosário had used the river for transport—and as a giant dumpster for their waste. In 1964, nearly sixty-six tons of human feces and a quarter million tons of urine entered the river every day.[1] As a consequence, Rosário was infamous for generating repeated outbreaks of typhoid.

The third-largest city in the country, Rosário was also home to major meat-processing and canning factories.[2] Canned products were sent to Europe. Corned beef was sent to grocery stores all over England. The manufacturing process at the canning facilities required heating up the tins, which were then cooled using river water. As a safeguard, the factories were supposed to chlorinate the polluted water, but the chlorination plant hadn't worked in over a year.

The cans that were set to be cooled were supposed to be watertight. No pollutant was ever to touch the interior contents. It worked most of the time, until one six-pound can went through the process with a small hole in the top. The untreated water entered the can. The large can, produced by a company called Fray Bentos, made it across the Atlantic to Aberdeen, Scotland, in May 1964.[3] From there it reached a grocery store on Union Street in the center of the city. Half of the meat from the can was put in the store window, the other half behind the cold-meat counter.

People in the city started falling sick with typhoid, and as the epidemic grew, so did the panic. The headlines were sensational, and news correspondents erroneously reported that people were dying in the streets. In fact, no one died during this outbreak, but more than five hundred people were afflicted, and it proved a serious public health crisis throughout the city. Ultimately, the source was determined to be the contaminated beef; the contamination spread to the store's meat slicer, which in turn came into contact with the other meat products sold there. The lead bacteriologist who solved the mystery was Dr. Ephraim "Andy" Anderson.

Discovering the origin of the typhoid outbreak brought Anderson immediate fame and recognition.[4] In 1965, he was the head of the Enteric Disease Lab within the Public Health Laboratory in Colindale, England. That year he also made a startling announcement. He had been researching various bacterial infections in calves and had found *Salmonella* that was resistant to two important antibiotics, ampicillin and chloramphenicol. This, he said, was the same strain of the bacteria that caused *Salmonella* poisoning in humans. Anderson then made a big leap—one that simultaneously irritated the agricultural lobby and the pharmaceutical companies.

He argued that the resistance in farm animals was due to the indiscriminate use of antibiotics in industrial food animal production, and this was likely going to result in increased resistance to antibiotics in humans.[5] The drug companies earned millions selling antibiotics to farmers, and farmers came to rely on the antibiotics to control disease among industrially raised animals.

But pressure from the public continued to mount, especially following a 1967 outbreak of *E. coli*-related gastroenteritis in West Lane Hospital in Middlesbrough. The bacterial infection proved resistant to ampicillin, streptomycin, tetracycline, chloramphenicol, kanamycin, and sulfonamide. Ten children died, and the news shook the country.[6] A BBC report linked the disease to the antibiotics in animal feed. Anderson's public statements about the risk of antibiotic resistance, though damning, had stopped short of drawing a definitive connection between the disease that had killed the children and the disease presenting in some animals. Both in the public and within the government, there were growing concerns about antibiotic resistance, and government wanted to protect public confidence in antibiotics. Responding to these pressures, the government quickly created a commission to investigate drug resistance in the livestock and agriculture sector.

But there was an immediate problem: whether or not to include Anderson on the commission. He was a well-known but prickly figure. The agriculture ministry knew of his strong stance on antibiotics in the farming sector and his insistence that farmers were culpable. The agricultural industry clearly wished that he would be blocked from the committee, and ultimately he was, much to the outrage of the scientific community. To assuage the critics, the committee announced that Michael Swann, another well-known though less controversial scientist would join them.[7]

Swann was a cell biologist and vice chancellor of the University of Edinburgh, and his name became synonymous with the committee. The Swann Committee was launched in the summer of 1968 and focused on antibiotic use in the agricultural sector. Its members divided the antibiotics into two categories: high-dose antibiotics that were used as therapeutics to treat infections and tackle outbreaks, and low-dose antibiotics that were used as growth promoters (in other words, the drugs that were used to increase meat production in the agricultural arena). The committee found no particular concerns with the practices governing the first category. It was the low-dose growth category on which the committee focused (and not on the drugs and doses used for treatment or infection control)—and the focus was limited to drugs that were important to human health. No attention was paid to drugs that were for exclusive veterinary or agricultural use. Their recommendations did not go as far as Anderson had been hoping for—much to his displeasure, the committee did not address the prophylactic use of antibiotics (in other words, using them as a *preventive* measure against possible disease). Ultimately, however, the committee made one bold recommendation—to ban penicillin and tetracycline as growth promoters.[8]

The Swann report was issued in November 1969. Six months later, a new government was in place, led by Prime Minister Edward Heath. The new regime implemented the recommendations, and penicillin and tetracycline were banned as growth promoters.

There was a major loophole. The drugs could still be prescribed by veterinarians who sided with the industrial farmers and had ties to the agricultural industry. These drugs could be used under the guise of prophylactic infection control—but, in fact, they were being used by the farmers as growth promoters. In light of the new regulations by the government of the United Kingdom, new committees were established to enforce the regulations, but the loopholes meant that these were not nearly as effective as the government would have hoped. In addition, the livestock industry was not required to share any information with the government regarding any evidence of resistance.

For all its failings, the Swann report was the first significant governmental effort to stop the use of certain antibiotics as growth promoters in animals. The world took notice. The Netherlands, Germany, and Czechoslovakia soon followed suit with their own laws.[9] In addition, over the next decade, the total volume of antibiotics given to livestock in the UK was cut in half. Over time, however, old practices slowly and gradually returned, and by 1978, the volume of drugs used in industrial livestock farming exceeded what had animated Anderson only a decade earlier.[10]

Similar regulatory policies around the world failed to generate political support, and nowhere as clearly as in the United States, where the FDA had adopted a strategy similar to that outlined in the Swann report in hopes of at least controlling the use of tetracycline and penicillin on America's industrial farms. The FDA's strategy was met with fierce resistance.[11] The list of opponents included some scientists and powerful lobbyists. But the strongest pushback came from the Animal Health Institute (AHI). AHI was created in 1941 with an interest in promoting industrial agriculture. With strong ties to US pharmaceutical companies engaged in the livestock and industrial sector, AHI had all the resources it needed to be an effective lobbyist against the FDA in Congress.

AHI took a risk and decided to settle the matter once and for all by commissioning a study that would definitively prove, beyond any doubt, that antibiotics as growth promoters were both useful for agriculture and essential for the economy. AHI found a relatively young clinical researcher by the name of Stuart Levy to undertake the study.[12]

Levy, who was based at Tufts University at the time, got to work immediately, enlisting farmers outside of Boston to participate in the research. His study was focused on chicken, where the hatching and growth time is significantly faster than in pigs or cows. Levy did a series of control experiments in which one group was given feed with antibiotics and one was given feed without. He then gathered up chicken poop from both groups to test the respective gut bacterias' resistance. In a short span of a few days, the gut bacteria of the birds getting the feed with the antibiotic started to change. The susceptible bugs were being killed by the antibiotics and the resistant bacteria were thriving.

In a few weeks, it got much worse. Now, the birds that had not been given the antibiotic at all started developing resistant bugs in their guts. And a few weeks later, *all* the birds had bacteria resistant to a whole host of antibiotics, including those that were not even in the feed.[13]

Levy, having proved the exact opposite of what the AHI had hoped for, published his findings in 1976. It was the most definitive study—anywhere in the world—that showed how adding antibiotics to animal feed resulted in antibiotic resistance traveling both horizontally between animals and vertically up the food chain. The scientific community was impressed and took note. Levy's academic star was ascendant, and he threw himself into further research and work, much of it trailblazing. He created the Alliance for the Prudent Use of Antibiotics (APUA)—which became a go-to institution for scientific evidence against the dangers of overuse of antibiotics. The scientific community was ecstatic about strong clinical and public health evidence linking antibiotic overuse and harm to human health.

Governments, politicians, regulators, and livestock industry were much less enthusiastic. The FDA regulations were not changing. The conservative governments of Margaret Thatcher in the United Kingdom and Ronald Reagan in the United States were not interested in new regulations. The Animal Health Institute, ignoring the study it had funded, continued to argue that there was little scientific evidence regarding any harm to humans. There were the rare cracks in corporate behavior—the McDonald's Corporation announced that it would only procure meat from suppliers that did not use any antibiotics that were important in human medicine.[14] But government action continued to lag, vacillating with the political fortunes of anti- and pro-regulatory politics.

While inertia and commercial interests continued to stall progress in the United States, discoveries from scientists like Levy in understanding how resistance spread, and from clinicians like Holmes on how existing antibiotics could be used more effectively, meant that not all was lost. Fresh ideas continued to come into view, including an ingenious plan, devised in Scandinavia, for making large amounts of data widely available in hopes of improving clinical and veterinary practices. This development would soon become a model for global change.

CHAPTER 19

THE NORWEGIAN SALMON

Tore Midtvedt was just a young boy when Norway fell under Nazi occupation.[1] Tore's father, Karsten Midtvedt, was an officer in the Norwegian navy and an inventor. He had visited Berlin in November 1938 to discuss his designs for a new radar antenna. The German industry was significantly more developed than the Norwegian one. Using his contacts within the military, he scheduled meetings to discuss his design. On November 8, Karsten Midtvedt witnessed Kristallnacht, the Germany-wide pogrom against German Jews. The violence and state-sanctioned lawlessness were heartbreaking and frightening. Karsten concluded that there was no way he could stay in Germany any longer or work with the Nazis. He returned to Norway the next day. Now as the war raged, Karsten knew that if the Nazis caught him, he would be imprisoned and quite possibly killed. The Midtvedt family, trying to protect Karsten, moved from one village to the next, evading bombs and soldiers, and they ultimately managed to elude the Nazis. Other members of the extended family were not as fortunate; several of young Midtvedt's aunts and uncles were sent to concentration camps.

Tore's first experience with antibiotics came right after the war, when his father, Karsten, suffered a fierce streptococcal infection that rapidly made its way through his body. Karsten's chances of survival were slim. Doctors injected him with penicillin every day for a week, and miraculously he survived, though it took time to appreciate the nature of the miracle. As he was being discharged from the naval hospital, the doctors told Karsten that he was the first Norwegian to be given 1 million Oxford units of penicillin (about half a gram in total), a unit of measurement devised by the Dunn team in Oxford during the early days of penicillin discovery. The first patient to receive penicillin, Albert Alexander, had been given a mere 200 units. Though Alexander improved initially, he did not survive; Karsten, on the other hand, was back on his feet after a week, fully fit.

By the time Tore Midtvedt finished medical school in the late 1950s, antibiotic resistance was a known phenomenon, though little was being done about it. Beginning his work at the Bacteriological Institute at the national hospital in Oslo, he was put in charge of standardizing the diagnostic discs used by clinicians to measure whether a particular clinical sample was susceptible or resistant to an antibiotic. Most of these were urinary samples to test urinary tract infections, largely from female patients at the hospital. Midtvedt's work mainly consisted of evaluating whether patients were resistant to the standard drugs, and if so, what the best alternative option was for their treatment.

As part of his job, in the early 1960s, Midtvedt was asked to evaluate ampicillin, a new drug produced by Astra, a Swedish pharmaceutical company. When he cultured the bacterial cells and tested ampicillin with the standardized diagnostic discs, the response was not what he had expected. Ampicillin was supposed to kill bacteria swiftly and uniformly. What Midtvedt saw was the exact opposite. The drug was sometimes effective and sometimes completely impotent. Midtvedt investigated whether the drug's potency depended on the acid levels of the bacterial sample. He also noticed that when Gram-negative *E. coli* was treated with the drug, it routinely became resistant to it. This response shouldn't happen with a new drug. Midtvedt repeated his experiments several times; the results were the same each time.

He wrote up his findings and sent one copy to a Norwegian scientific journal and the other copy to Astra. He got a call from Astra immediately, and the head of its local research department soon took him out to dinner. Over drinks, the head of the research department asked if Midtvedt would be interested in working for Astra. The company would be glad to provide him with additional resources for his studies. The catch was that he'd have to withdraw his findings from publishing and wait until Astra endorsed them.

The same instincts and ethics that had sent his father back to Norway all those years before guided Midtvedt forward. He did not accept Astra's offer, and his relationship with the company would remain strained for years, particularly during the time he was studying in Sweden, Astra's home base, pursuing his PhD at the prestigious Karolinska Institutet.

By the early 1970s, and with a PhD under his belt, Midtvedt was back in Oslo at the Oslo University Hospital (also known as Rikshospitalet, or RIKS). Once again, he was testing bacterial samples for susceptibility and resistance to antibiotics. Not only was the process painstakingly slow, but, worse, the accumulated information stayed largely in the lab. There was no broader system within the country for monitoring a failing drug and warning others that it was no longer effective.

As it happened, a huge IBM computer had been installed at the hospital, one that required punch cards to store data. Midtvedt had an idea: What if he could use the machine to store his samples and track the results? He talked to the head of the computer wing, who was intrigued. With the help of a young student, Midtvedt gathered clinical samples from all over the hospital and tested each one against thirteen different drugs available there. He labeled the clinical samples as sensitive, relatively sensitive, relatively resistant, or resistant to each drug. In addition, Midtvedt recorded the MIC levels (the minimum dosage required) to kill the infection-causing bacteria.[2] Midtvedt then sent the data from his lab to his student, who would enter it into the computer. Every day, dozens of records made their way to the computer. The team worked nonstop, dedicated to the cause of minimizing the spread of drug-resistant infection, and they made the data freely available to anyone at RIKS. By doing so, they could see both the trends of resistant infections and the dosage required for a drug to be effective. By 1980, a stunning fifty-five thousand samples had been logged into the computer.

Years after the project started, after it had yielded a treasure trove of information, Midtvedt got a call from the head of the computer department: there had been an issue with the computer, and all of his data was gone.

Midtvedt was devastated. He had put his heart and soul into the project, which was well ahead of its time. No other country was conducting data collection and organization that came anywhere close to what he had been doing. The applications had been immediate, and the consequences of this loss were going to be profound.

Disconsolate, Midtvedt asked the head of the computer department what to do with the punch cards that he had collected over the years. He was largely ignorant when it came to the inner workings of the computer, but the meticulous researcher had held on to his data. "Punch cards?" the head asked. "You mean you have the original punch cards?" Midtvedt said that of course he did: every single one of them. The head of the computer department couldn't contain his excitement. If true—if Midtvedt had kept all those cards—then the entire data set could be re-created, provided the department had the people to do it.

Midtvedt couldn't believe what he had been told. And yes, the same determination that had gathered the original data could be tapped to re-create the computer archive. Over the course of several months, Midtvedt and his students rebuilt the entire database and restored the majority of records.

By the early 1980s, Midtvedt's program of data collection, analysis, and information sharing was being used at RIKS and other facilities across Norway. As a result, a map of antibiotic resistance in the country was emerging, and it was going to play a major role in shaping the country's policy on controlling antibiotics prescription in the decades to come.

Midtvedt's home country of Norway has become synonymous with quality salmon all around the world;[3] much of the available salmon comes from Norwegian fish farms. The grocery store near my home in the suburbs of Boston sells it year round. Midtvedt, it turns out, had a role to play in the growth of the salmon industry. The fish in grocery stores all across America and in other parts of the world is there thanks to a vaccine that was developed to protect the fish from yet another threat of rising resistance due to industrial-use antibiotics.

The massive growth that we've seen in Norway's salmon industry over the years took off in the 1980s with the advent of modern fish farms. A typical farm consists of several large circular cages placed side by side in the sea. Some cages are only ten meters in diameter, but others can be as large as half a soccer field. They can be as deep as they are wide. These cages contain nets that can hold up to one hundred thousand salmon. Some even larger cages can hold nearly a quarter of a million salmon.

As the farming sector expanded, exports increased, but so did the spread of a devastating fish disease called furunculosis. To protect their livelihoods and their salmon, the farmers in Norway started using antibiotics as prophylaxis by adding it directly to the feed for the fish.

As the scale of the fish farming industry grew, so did the use of antibiotics. By the late 1980s, demand was such that the farmers were using cement mixers to mix antibiotics into the fish food. Midtvedt grew concerned. His years of research and his ever more detailed database told him that this method of farming could greatly affect the waterways, the environment, and the public in general.

By the mid-1980s he started writing about it publicly. And after obtaining data from the relevant government departments, he found that humans in Norway were prescribed twenty-four tons of antibiotics per year. Salmon consumed forty-eight tons. The fact that twice as many tons of antibiotics were being given to the Norwegian salmon than to all the people in Norway was alarming. That the "antibiotic-enhanced" fish food was being dumped directly into fish farms, and from there could get into the surrounding water, was even more so. Midtvedt pushed for some response from the Norwegian government, hoping it would enact regulations to protect environmental and public health, but fish farming was a lucrative and highly profitable business and a major source of tax revenue. He was met with fierce resistance by the fish farming lobby, and by veterinarians who had strong ties to the fish farmers.

In 1989, Midtvedt and other like-minded scientists got a big break from an unlikely source, the Norwegian State Broadcasting Corporation, or NRK. A local NRK station made a documentary about antibiotics and fish farming. The production team managed to take video cameras to the bottom of the salmon farms and found that the bottom surface of the farms had turned black from the presence of excess antibiotics. The journalists then took samples from water streams that were miles away from the salmon farms and found that the fish there also contained high levels of antibiotics. That wasn't all. They found evidence of antibiotics in birds that were feeding on the fish or on antibiotic-laced fish food that floated to the surface of the water.

The movie was explosive. It was shown only once on public television in Norway. The leaders of the industrial fish farms were incensed. Any threat to their businesses, they argued, was a threat to the fish economy and, consequently, a threat to the country's economy. The government told NRK not to show the movie again. The producer received anonymous bomb threats. NRK complied, and the movie was essentially banned.[4]

Yet the tide was turning. News of the movie, new data about resistance, and rising public awareness created a demand for action that was more than the farming industry could handle. The farmers realized that the more strenuously they strove to shut down inquiries into their practices, in hopes of not appearing to be the villains, the more they were, in fact, labeled as such.

They were saved by science. Around the time they could no longer quiet the upset over the extensive use of antibiotics, there was promising news coming in regarding a new salmon vaccine that could prevent furunculosis and decrease the need for prophylactic antibiotics. The vaccine—administered by injection into the abdomen of the fish by an automated process—was a godsend for the farmers, the industry, and the nation's economy. By 1994, vaccination had become routine in Norway, and the use of antibiotics plummeted.[5]

The Norwegian example would become a staple in conversations about antibiotics and meat production, and Midtvedt was hailed as the scientist who made the public aware of the issue. He is now widely recognized as an instrumental figure in creating a model of surveillance that was based on mountains of collected, tabulated, and carefully analyzed data. He built a coalition of like-minded scientists, one researcher at a time. And at a time when the pharmaceutical industry, let alone his own government, was unwilling to listen to him or pay any heed to his data, he persisted and pushed for changes in policy and practice. The evidence was on Midtvedt's side and eventually impossible for the government to ignore. His dogged attitude paid off. And in 2018, for his repeat advances in microbiology and for his services to the country, an eighty-four-year-old Tore Midtvedt was given Norway's highest civilian award, the Royal Norwegian Order of Saint Olav 1st Class.[6]

CHAPTER 20

CLOSER TO SYDNEY THAN TO PERTH

In 1992, a team of researchers in a caravan of Toyota 4x4 Land Cruisers headed to remote indigenous settlements in the far north of Western Australia.[1] Though the team was from Perth, the sites they were visiting were well over two thousand miles away. Some of the sites were farther from Perth than Perth is from Sydney, across the country on the east coast. Since 1957, Perth had been the home of Warren Grubb, a microbiologist who had grown up in New South Wales on the east coast. Grubb was now in charge of a team of scientists at Curtin University, and he was also on the cusp of a major breakthrough.

Grubb's team had been interested in staph infections for some time, which is in part why the Health Department of Western Australia had awarded them a grant to screen people in remote communities for methicillin-resistant staph aureus (MRSA). While there was some information on the level of MRSA among people coming to Perth hospitals, there was little knowledge about those who lived in the far north and eastern parts of the state. Health personnel in Western Australia were anxious to avoid having MRSA become an endemic problem in their hospitals as it had in the country's east-coast hospitals, where these strains of staph were consistently causing infections.

So far, Western Australia had been spared, largely because of its isolation. WA, as it is referred to, is the largest state in the country. It covers about a million square miles, roughly the size of Texas and Alaska combined. The state capital of Perth is one of the most isolated cities in the world. The nearest city of any comparable size, Adelaide, is over seventeen hundred miles away. And while the population of WA is about 2.6 million, nearly 85 percent of the population lives in Perth and its suburbs.

Grubb's lab was a hub for MRSA screening. The expertise was a result of a very successful partnership between Grubb's lab and a microbiologist at the Royal Perth Hospital, Dr. John Pearman. The State Health Department had mandated that all MRSA strains were to be sent to Grubb's typing laboratory. The Royal Perth Hospital had provided MRSA samples to Grubb and his team so they could screen and type them, then compile the information for infection control. The idea was to create a MRSA database so that strains could be identified and tracked in the hopes of preventing them from spreading into WA hospitals.

Over the years, Grubb's team had built one of the best such MRSA databases in the country, and, as a result, they were routinely relied on for help in identifying different types of MRSA. They had even discovered a strain that hadn't been seen in Australia before, which matched a type that they had received from Houston, Texas, half a world away. They traced that particular strain back to two sailors from Houston whose ship had docked in the southern part of WA, and, upon return to the United States, their infection was found to be MRSA.

But something odd started to happen. The team was receiving MRSA samples taken from patients in the remote Kimberley region in the far north of the state. These strains did not look anything like what was in their database. It was a mystery that the team couldn't fully pin down. At least not in Perth. They needed to go far into the interior of the state, visit the isolated communities, and learn more about the strains.

Grubb was concerned that the remote communities might not

participate sufficiently for his team to reach conclusive findings. Collecting patient histories was a critical piece in solving this puzzle, and that required full participation and trust. Luckily, Grubb secured some crucial support from one of the biggest names in the field of aboriginal health, Dr. Michael Gracey.

Gracey was a highly regarded clinician who had spent his career studying the native aboriginal populations in WA. In 1971, Gracey traveled to Kimberley for a monthlong trip to understand the health challenges in remote communities. What he saw there frustrated him. Childhood diarrhea and malnutrition were rampant. So over the next two decades, Gracey conducted extensive research to understand the socioeconomic factors and clinical challenges that were driving the health gap between white Australians and native aborigine populations.[2]

Gracey flew to Kimberly ahead of Grubb, enlisted a couple of local people to help with his fieldwork, and ensured that Grubb would be able to collect samples as well as record patient histories. Without Gracey's help, Grubb's team would never have been able to gain access to the communities, and without that access, their discovery would never have been so timely.

The team collected nasal, throat, and skin swabs from the communities. But sampling was only the beginning. There remained the challenge of how to go about processing them. There was no obvious way the team could successfully get these samples all the way back to Perth. There was, however, a unique Australian solution available. The team enlisted the help of the Royal Flying Doctors (RFD), a highly valued institution in Australia.[3]

Founded by the Reverend John Flynn about one hundred years ago in Queensland, the flying service was the first air ambulance in the world, introduced at a time when air travel was reserved for a privileged few. But the fusion of two innovations at the time, radio and airplanes, created a high-tech solution to the health crises that were all the more difficult to solve, given Australia's terrain and remote communities. By the time Grubb and his team had to call upon it, the RFD was a valued and renowned service that was about to play yet another key role in advancing the country's health.

When the team got back to Perth and analyzed the samples from the communities, they found that these matched the MRSA strains that were previously coming to their lab from parts of Kimberley, and they were different from the hospital strains in Perth. These strains, from Kimberley and other locations that Grubb and his team screened, had not been reported anywhere else in the world and were different genetically from all previously reported MRSA. Their plasmids and resistance patterns were completely separate from the strains that Grubb had seen from other parts of Australia and on a global level. More peculiar was the fact that the people from whom these new MRSA samples had come had no prior history of hospitalization. MRSA, up until this time, was known to be a hospital-acquired infection—not known to be present in the communities with no history of hospitalization.

The new strains of MRSA raised another question: Was this also happening in other isolated indigenous communities? To find out required testing on a scale that had never been done in WA. Grubb wrote up a proposal and received a grant from the Australian National Health and Medical Research Council. For the next seven years, with assistance from the WA Health Department and the Royal Flying Doctors, Grubb's team crisscrossed the state screening people in other remote indigenous communities far from the Kimberley region—and, in between trips, they'd stop to analyze the results in Perth. They traveled to the Warburton Ranges in the eastern part of the state, to Pilbara in the heart of the state, and on and on, taking samples from communities small and large. Some communities consisted of just forty people, others of a few hundred. Some people were sampled multiple times over the period of several years to see how MRSA was emerging in their environments. Again and again, Grubb's team was helped along the way by doctors like Gracey, who had a long history of working with indigenous peoples.

No matter where they went, Grubb and his team would take note of patient histories and, crucially, whether any antibiotics were being used in the communities. Piece by piece a picture came into view, one that defied the world's previous understanding of MRSA. Carefully typing the strains, and connecting them with patient histories, the team fundamentally altered the existing understanding of how resistance emerges.

Their findings were ominous: they found a twin to hospital-acquired MRSA, which they abbreviated CA, for "community acquired." Much to the horror of scientists and public health professionals around the globe, Grubb had proved conclusively that MRSA did not have to come exclusively from hospitals.[4]

It could come from a community.

CHAPTER 21

A CLASSLESS PROBLEM

Septran, a combination of the two antibiotics trimethoprim and sulfamethoxazole, was a staple in my childhood home. If I had a fever, I got a tablespoon of the pink syrup. If my throat itched, I got a teaspoon. I never minded it—it was sweet, a combination of bubble gum flavor and other additives—and it often meant that I would miss school. I outgrew the flavored syrup, but the drug remained, always there, in the kitchen, in the second cabinet from the left. Septran, I learned, was not just for kids—nor was it only available in a pink syrup. There was a tablet form as well—in a blue blister pack with white oval pills.

There was never a question of needing a prescription. Anyone could go and buy it from the pharmacy, if they could afford it. We were a middle-class family, and a marker of that was a house well stocked with drugs.

Three decades later, little has changed in Pakistan. The pharmacies in my old neighborhood now sell brand names and generics, international brands and local ones. The price of drugs varies, and so does the quality, but the ability to buy them without a prescription remains the same.[1]

Local doctors, including many in my own family, are part of the problem of prescribing antibiotics when they are going to create more harm than good. Whether the patient presents with a fever, sore throat, or toothache, the family doctor is ready to suggest a course of antibiotics, sometimes simply over the phone. No tests are ever done to determine whether the infection is, in fact, due to a bacteria. Worse, patients often show up at the pharmacy to get antibiotics themselves, without any prescription at all, similar to what Dr. Holmes observed in the Philippines. The pharmacists are equally eager to make the sale.

In the 1960s, antibiotics started to become widely available in Pakistan and in a number of other developing countries. The patents of branded drugs were expiring and new generics were coming to market. Most of the drugs in low- and middle-income countries were imported at that time, but by the 1970s, the local pharmaceutical companies in a few developing countries were increasing in size and in production rates. In India, the pharmaceutical industry got a big boost in 1970 from new patent laws;[2] the government made it possible for local companies to produce drugs that were still protected by patents, by defining patents on the process, and not on the product. As long as Indian companies were using a different process, they could make something identical to what international companies were making.

Indian companies started to "reverse engineer" in order to come up with processes by which they could make products identical to what was available in more affluent countries.[3] The change in the patent laws made it very difficult for international companies to sue Indian manufacturers for infringement. The boom in the Indian industry created a bigger supply, making the drugs more affordable and accessible not just in India.[4] In other parts of the world, local and even international pharmaceutical companies eager to increase their sales often pushed back on prescription drug laws.

The ubiquity of antibiotics, relative affordability, weak regulatory laws, and lack of enforcement has meant that for many middle-class households, a cabinet well stocked with antibiotics is the norm.[5]

Over time, we started noticing that we had to double the dose of our own antibiotics. Instead of one spoonful of Septran, it had to be two. This problem was not unique to my family—residents across the city, even those who had far fewer means and resources than we did, were facing a similar challenge.

Not far from our home in Islamabad is an urban slum called France Colony. The name of the slum is an accident of history. It is the land that was used to house the French embassy, before it moved out to a more secure location. Now, the majority of people living in France Colony are the very opposite of descendants of an affluent Western European nation.[6]

Though living in the center of the city very near to one of its commercial hubs, the residents are part of Pakistan's marginalized Christian population and are often targeted, persecuted, and treated poorly by the government and the majority Muslim citizens of Islamabad. In France Colony, the sewage and drinking water routinely mix; it is an epicenter of diseases. Impoverished parents of young children who face threatening infections first do what my family always did—they start with home remedies and medicines like Septran that don't require a prescription. When those methods fail, the children often end up in the ward of a nearby hospital that is overflowing with patients and often lacks the resources needed to help them.

A longtime resident of France Colony, Sahiba* works backbreaking hours as a maid in the house of one of the city's elite. When I met Sahiba, she had recently lost her baby to an infection that the doctors at her local hospital couldn't effectively treat with

---

*Name changed to protect identity.

any of the antibiotics they tried. Implicitly and explicitly, the doctors blamed Sahiba for exposing the baby to an unhygienic environment. Sahiba could do very little about the neighborhood in which she was raising her children, and the pharmacies in the nicer neighborhoods surrounding it continued to sell a steady stream of antibiotics that found their way into medicine cabinets across Islamabad.

Sahiba's story is not unique: the resistance rates in the country are staggering, and they ultimately affect everyone.

When I was growing up, I never realized that taking a spoonful of Septran for a runny nose was a problem. I thought I was being proactive in arresting the development of my illness before it became worse. Drug resistance was never mentioned to me, and the drugs were available and seemed to work. Or so I thought. I was of the right sect, the right ethnicity; I lived in a good neighborhood and had access to clean water and sanitation. The same system that failed Sahiba was generous to me and everyone around me. Infections of the sort that threaten Sahiba and her children were lamentable, but inevitable, given the dire levels of poverty in the area.

Sahiba and her fellow residents of the France Colony were often blamed by the elite for the problem of spreading disease— but it was the social system that created abject poverty and a dysfunctional government that failed to prioritize hygiene and public welfare.

I didn't realize at the time that my family and I were also guilty, through our drug-consumption habits. We now know that we are all part of the problem. And the longer we continue to blame the impoverished for the spread of infections around the world, dismissing the ways in which middle-class and wealthy people contribute to antibiotic resistance, the longer it will take to address the problem. We would all be better off if we realized that sometimes it isn't the patient's fault—sometimes it's the fault of the pharmaceutical companies and the governments that should be regulating them.

In another slum in the Pakistani city of Lahore, Sadiq* was insulted by the report issued by the local health authorities blaming his father, Aslam, for his own death. His father had died after he took the only recreational drug that he and his friends could afford: cough syrup. Aslam had been doing this for years. Every weekend, Aslam and a group of his friends, all of them from the same slum, would get together and consume one bottle of cough syrup each. This was their weekly treat—a simple meal of greasy street food, a bottle of a cough syrup that acted as a mild hallucinogen if you took enough of it, and occasionally a card game. Then they would return home, sleep off the effects, and go back to work the next day.

Except this time, Sadiq's father came home, had dinner, and never woke up. Nor did any of his friends. More than a dozen people died—all from consuming the exact same cough syrup. Coming on the heels of another scandal related to the quality of cardiovascular drugs that had claimed the lives of more than two hundred people in Pakistan,[7] government officials were fearful of more negative publicity, so they swiftly blamed the consumers of the cough syrup. Sadiq was angered by this—the government refused to take any responsibility and announced that Aslam and his friends were all to blame for their own demise.

They were addicts, said the provincial health chief. Some went even further and called them the scum of the Earth, no better than grave dwellers—lowlifes who were not fully human. A week later, the same problem of a toxic cough syrup emerged in a neighboring city—this time claiming the lives of more than thirty people.[8] Again, the government, supported by the local pharmaceutical industry, dug in its heels and claimed that it had lab reports confirming that there was nothing wrong with the syrup.

---

*Name and that of Aslam below changed to protect identity.

Sadiq challenged the government to publish proof that the deaths weren't linked to tainted syrup. A couple of newspapers took notice of his challenge, but other events soon crowded him out, and the world moved on. The lax controls at pharmaceutical companies continued.

Near the slum in Lahore stands the Gulab Devi Chest Hospital. That it's named for a Hindu philanthropist has been a thorn in the side of religious zealots. That it's home to one of the largest TB hospitals in the country has ensured its importance in the global confrontation with drug-resistant bacteria. With close to fifteen hundred beds, the hospital is in a continuous state of chaos. Here, doctors treat numerous TB patients, including many who suffer from MDR (or multi-drug-resistant) TB. Technically, MDR TB refers to infections that no longer respond to first-line drugs—rifampicin and isoniazid are the two that are most commonly used. There is a correlation between MDR TB and poverty, for most of the patients who come to Gulab Devi are indeed impoverished. They fit the stereotypical definition of TB in Pakistan: that it is a disease of the poor.

Kulsoom Bibi* was one such patient, and she had been coming to Gulab Devi as an outpatient for over a year. She came from the same area as Sadiq, and, like Sadiq, she was poor. She was also treated poorly. Her doctor at the hospital would often speak to her harshly, telling Kulsoom that it was her fault that her infection was not responding to the drugs he prescribed. She must not have adhered to the treatment regime she had been given.

Kulsoom Bibi insisted that the opposite was true, that she had been diligent in following instructions. She took great pride in the little education she had, and she told the doctor that she knew how to count and tell time, and that she knew she had taken all her pills on time. Unconvinced, uninterested, already turning to

---

*Name changed to protect identity.

the next patient, the doctor ended the appointment. The verdict remained. It's an utterly common one.

The argument about adherence goes something like this. Those prescribed an antibiotic are supposed to take their drugs for a set and strict period of time. This period is based on experimental and clinical studies that have shown how long patients should take the drugs to kill all of the infectious bacteria. Because some bacteria may take longer to die than others, taking the drugs for the full period is critical to recovery. Ending the regimen before that period means that some bacteria can and likely will survive—and could become resistant.

The active part of any drug is called the active pharmaceutical ingredient, or API. Inside the body, different drugs release their API at different rates. However, if a drug does not release enough API or does not even have enough API, the bacteria would be exposed to an insufficient amount of the antibiotic, an amount that would not effectively kill the whole population. The bacteria that survive can best resist the antibiotic: when the antibiotic kills off most of the competition, the resistant bacteria are free to proliferate, and in some cases even mutate further, moving from partial to complete resistance.

The chance of selection of resistant bacteria is the reason why doctors insist that patients take their drugs for the entire prescribed period. This is why Kulsoom's doctor was irate. She ruled that Kulsoom was a bad patient who made her once manageable disease a lot more difficult to manage. Worse, by making her disease drug resistant, she was going to expose the people around her to an infection growing ever more powerful.

But the argument should be looked at another way. What if the drug was only 50 percent pure? Even if the patient took a full course of the antibiotic, doing so would be like taking it for half the prescribed period. The patient may be fully compliant, but the system that was supposed to provide the care had failed the patient. Like cough syrup killing the poor in Lahore, what if the drug and those creating and distributing it were the real problems?

Drugs that contain a dose less than what is written on the package, or drugs that degrade because they are poorly manufactured or stored in high heat when they are supposed to be refrigerated, continue to be a major problem facing many countries. The companies operating in these countries are more likely to cut corners in quality assurance and produce substandard drugs due to a lack of tough regulations and enforcement. The global supply chain also means that a drug sold in Kenya could be made in India with ingredients from China, further increasing the likelihood that any bad actor along the way could create problems for the unsuspecting consumer.[9] Global politics, aid dependency of poorer nations, and new forms of colonialism help suppress the problem. Governments that are receiving substantial funds from China to undertake huge infrastructural development projects, for example, are reluctant to blame state-owned Chinese pharmaceutical companies. Often, in poor countries, the national health resources are stretched so thin that they are unable to conduct any tests to ensure that the country's drug supply meets the international standards of quality and safety.

Conservative estimates suggest that at least 10 percent of all drugs in low- and middle-income countries are of poor quality.[10] In some countries, as much as one-third of all available drugs, both prescription and over the counter, are of poor quality.[11] The numbers of patients who die every year as a consequence is in the hundreds of thousands around the world. And antibiotics are among the most common substandard or outright fake drugs out there today. Until recently, it wasn't clear if these drugs could cause resistance. Now we know that they can.

Zohar Weinstein was busy pursuing her PhD in my research lab at Boston University, and about a year into her work she found something peculiar. She was studying whether new combinations of existing antibiotics could overcome resistance, and she was investigating a particular drug called rifampicin. Rifampicin had been discovered during the heyday of antibiotics research, and, like so many of the drugs, it came from a soil sample.

In rifampicin's case, the sample had come from the French Riviera in 1957 and was sent to Lepetit Pharmaceuticals research lab in Milan for analysis. Two scientists at the lab, Piero Sensi and Maria Teresa Timbal, identified a new bacterium in the soil sample from the French Riviera that produced a new class of antibiotic molecules that eventually came to be known as rifampicin. It took about a decade for the drug to come to the market, first in Italy in 1968, and after receiving FDA approval in 1971, it came to the United States. It soon became a frontline drug against TB.[12]

In the course of her studies, Weinstein noticed that the drug degrades rather easily into a compound called rifampicin quinone—a molecule that is considered an impurity by pharmaceutical companies and regulatory agencies. Put bluntly, rifampicin that has rifampicin quinone in it should not be on the market, even if it contains only a small amount of the impurity. This reaction from rifampicin to rifampicin quinone can happen under certain environmental conditions, but there are additives (such as ascorbic acid) to make sure it stays as rifampicin.

The ease by which rifampicin became rifampicin quinone intrigued Weinstein, and it bothered her. What, she wondered, was the effect of this on bacterial resistance? She systematically exposed bacteria to both pure rifampicin and the impure substance. She also simulated the adherence problem by giving the bacteria a lower dose than a doctor would likely prescribe.

Most impurities wouldn't kill the bacteria, but they also wouldn't cause resistance. It was like giving bacteria a placebo. But with rifampicin quinone, Weinstein saw something shocking. The impurity not only *caused* resistance, it caused it much faster than even lower doses of rifampicin. Administering the drug with the impurity was even worse than the problem of adherence.

Weinstein repeated her experiments dozens of times. Every time, she confirmed her results.

She wasn't done. Weinstein reached out to Eric Rubin for advice. Rubin, a TB expert at the Harvard School of Public Health, also found the results troubling and encouraged Weinstein to test her hypothesis using another bacteria called *Mycobacterium smegmatis*, which was a more realistic lab model organism to study TB since the actual TB pathogen grows very slowly in the lab and requires extensive safety protocols to be in place. Weinstein started the painstaking experiments once again—separating various cultures, keeping track of each individual dose, and then seeing how the bacteria responded to therapeutic concentrations of rifampicin, concentrations lower than the therapeutic level, and exposure to just the impurity. Here, the results were even more surprising. Resistance to rifampicin developed much faster after exposure to the impurity than when the bacteria were simply exposed to a lower dose.[13]

Weinstein had discovered a direct link between drug impurity and resistance. She took the next step and conducted experiments to find out what was happening at the genetic level. Were there some new mutations—changes at the molecular level—that were enabling the bacteria to become resistant? Once again, she found new mutations, some that had never been reported before.

The final question was what happened when bacteria that fed on the impure drug were hit with the pure, potent drug. Weinstein selected the bacteria that had been exposed to the impurity and had become resistant. She added the pure rifampicin and kept adding it, but the bacteria was permanently resistant. No amount of rifampicin tested could kill it.

This was another revelation. Resistance occurred not just because doctors overprescribed antibiotics; it occurred not just because patients failed to adhere to their full treatment regime, or were consuming a substandard drug—one that had been adulterated with impurities and degradation products. It was no longer just about drugs becoming part of industrial food production. Resistance was clearly also due to the presence of substandard drugs and because of the failure of the regulatory systems to prevent impurities from getting to market.

CHAPTER 22

THE STUBBORN WOUNDS OF WAR

In November 2017, I agreed to join colleagues at the American University of Beirut for a two-day meeting. I had the chance to be part of an international team that was aiming for a $5 million grant to fund research into antibiotic resistance. The request for proposals had come from the Medical Research Council in the UK, and they expected that any project that would get the grant would be one with an ambitious scope. I was part of the Lebanese-led team, and we were finalists for the grant. The team represented an array of technical expertise and included researchers from France, Sweden, Yemen, Jordan, the Netherlands, and of course England.

The problem we wanted to study was persistent and drug-resistant infection among patients arriving for treatment in Lebanon from other Middle Eastern countries, particularly Syria and Iraq. I knew that the issue was acute, but I didn't expect that the central hypothesis of my gathered colleagues was that it traced back to an event that happened nearly fifteen years before: the American-led war in Iraq.

Soon after the 2003 invasion and occupation of Iraq, the doctors at the US Army field hospitals started noticing high levels of infection related to an opportunistic bacteria, *Acinetobacter baumannii*.

It's considered opportunistic because it doesn't cause a disease on its own, but if there is an existing infection—pneumonia or an infected wound, for example—it thrives.[1] The bacteria is found everywhere—in streams and soil, on the walls of hospitals, and on the skin of its patients.[2] Once established, it takes off.

Though it was persistent, the doctors knew that the bacteria could be controlled.

But not this time. The problem facing the US field hospitals was not the widespread prevalence of the bacteria; what bothered the military doctors was that this Gram-negative bacteria was showing serious resistance to a large number of the best antibiotics available.[3] Initially, they encountered just a few such cases, but the numbers were creeping up. Over a six-year period, from 2003 to 2009, nearly thirty-three hundred US troops were treated for drug-resistant *Acinetobacter baumannii*. Worse, the veterans were bringing it back to the US hospitals, like Walter Reed National Medical Center, where they were often treated after their stays in field hospitals in Iraq.[4]

During the protracted war, *Acinetobacter* came to be seen as a major threat to American troops in Iraq. It even earned a new name: Iraqibacter. The problem was so acute that in 2010 Congress convened a special hearing on the issue.[5] But then the US military slowly left Iraq, and as the numbers of troops decreased, so did the military field operations. These days, drug resistant *Acinetobacter* is no longer high on the list of concerns threatening the US military, but the problem persists for the local population of the region.[6]

Did the US invasion of Iraq cause the development and rise of Iraqibacter, and, more broadly, was there a link to America's wars in the Gulf in general and the 2003 conflict in particular that set drug-resistant *Acinetobacter* on the world?

Ghassan Abu-Sittah and his colleagues in Lebanon believe that the answer is a resounding yes.[7] Abu-Sittah is one of the region's leading plastic surgeons and has seen hundreds of patients suffering from the traumas of war. Born to a Palestinian father who received his medical degree in Cairo, as many Palestinians did before the Camp David Accords, Abu-Sittah received his own medical training in the UK. First trained in Glasgow and then in London, Abu-Sittah had been traveling to conflict zones in the Middle East since the early 1990s, when he worked in Iraq and South Lebanon.

In 2009, he was recruited by the American University of Beirut to head its Department of Plastic Surgery at the university hospital. Back in Beirut, Abu-Sittah saw an ever greater number of patients suffering from persistent infections. When he cultured these infections, they'd come back positive for Iraqibacter.

As Abu-Sittah started his work, he met another researcher at the university, Dr. Souha Kanj, now head of Infectious Diseases at the American University of Beirut's Medical Center. Dr. Kanj is no stranger to conflict and war.[8] She was a medical student at the French University in Lebanon when the nation's fifteen-year-long civil war flared up, making it impossible for her to continue her studies. She went abroad for a year, studying in Bordeaux, returning to Lebanon when, she hoped, things would've calmed down. They hadn't. She ended up enrolling at the American University of Beirut (AUB), which was closer to her home, and reduced the risks she would face at the checkpoints all over Beirut. After she completed her degree, Kanj trained at the Duke Medical Center and became a pioneer in infectious diseases and solid organ transplants.

In 1998, she and her husband moved back to Lebanon. During her first years back, the infections that Kanj saw were similar to those being reported in other parts of the world. But in 2006, things took a sharp turn. Israel invaded Lebanon again, and the war destroyed buildings, bridges, and infrastructure. At the same time, the patients who were coming to the infectious disease unit at AUB started showing signs of severe resistance to nearly all antibiotics. With mounting concern, Kanj ordered that every single sample from a patient who came to her ward be cultured rigorously, and by 2007 she and her team recognized that they were dealing with an outbreak of *Acinetobacter baumannii* that was resistant to everything in their arsenal.

Kanj wanted to dig further. Had anything like this ever happened before? She asked around, especially in the microbiology lab at the university. Most of the staff replied that this was new to them as well, except a couple of older microbiologists. They looked at the culture and nodded. They had seen this before, in the years 1975 to 1990, during the civil war that had devastated Lebanon. Kanj had never heard of this, and she asked if they had ever published their findings. Given the crisis, they hadn't had the time. She now had a hunch—something linked war to *Acinetobacter* resistance.

Kanj had heard of Abu-Sittah's work in trauma, and Abu-Sittah was familiar with Kanj's expertise in infectious diseases. Taking time away from their clinical practice wasn't easy, but there was something special at play here. They hit it off during their first meeting and knew it was time to work together—and to find a group of people who would serve as the best model to test whether drug-resistant *Acinetobacter baumannii* had something to do with conflict.

They did not have to look too long, or too far. Among the doctors' patient population were a substantial number of Iraqis. At the time, they routinely sought treatment in Lebanon, Jordan, or Turkey for any serious condition. The reason was simple enough. The health-care system in Iraq, once one of the better systems in the Middle East, had collapsed completely first due to sanctions after the first Gulf War of the 1990s and then the US invasion of Iraq starting in 2003.[9] Patients who could afford it, or were influential enough and could get the Iraqi government to pay for it, went to facilities in nearby nations. Abu-Sittah remembers one particular patient who had intercepted a suicide bomber and been badly injured. The trauma was severe, and the patient's bones became infected. Typical frontline antibiotics were just not working. Abu-Sittah requested a blood culture, and there it was, *Acinetobacter baumannii*.

Abu-Sittah and his colleagues would see patients suffering from gunshot wounds and survivors of bombings and traffic incidents. Because of severe trauma, many of them had developed osteomyelitis, an infection of the bone. The infection would give rise to colonization by drug-resistant Iraqibacter—and, with few drugs available, the prognosis for the patients did not look good.

To figure out what was behind this rise in drug-resistant Iraqibacter among Iraqi patients, Abu-Sittah teamed up with another colleague, an anthropologist named Omar al-Dewachi. Dewachi's area of expertise included studying the Iraqi public health system, especially its recent history. He had lived through it, during the first Gulf War.[10]

Starting in the early 1990s, with the onslaught of America's first invasion of Iraq, the Iraqi public health system had begun to collapse. War was followed by sanctions, which meant that few nations of any means were willing to do business in Iraq. The impact on Saddam Hussein's inner circle was marginal; for the wealthy and the privileged, there were ways to secure quality care. But the effect on Iraq's general population and public hospitals was severe.[11]

Omar al-Dewachi had finished his medical training the year the war started and witnessed it as a doctor in Baghdad's main hospital. Supplies were limited, and they were degrading. He remembers that even the masks that the surgeons wore to avoid infections had to be recycled and reused until they disintegrated. Often there were no masks whatsoever.

The first Gulf War had left Iraq's health-care system in shambles. The invasion in 2003 broke it completely. Doctors weren't getting paid and did not have the necessary equipment to do even basic work. Many left or escaped Iraq. Dewachi was one of them. He ended up in Beirut and then at Harvard, where he began work toward his PhD. In the midst of his studies, Dewachi traveled to Canada, where he was told that he could not reenter the United States on his Iraqi passport. Issued during the Saddam regime, his passport was no longer a valid document.

Stranded, Dewachi met and started working with a medical anthropologist in Canada, Dr. Vinh-Kim Nguyen.[12] Nguyen's interests were uniquely international. His parents were originally from Vietnam and Switzerland, and he had grown up in England and then moved to Canada. Over the course of his studies and research, he had become deeply interested in the global response to HIV, and what it had meant for those suffering from AIDS in Africa. Dewachi's interests in the issues of access to care, trauma, and the wounds of war resonated with Nguyen, who became his host while he figured out how to get back to the United States. The two became close friends and academic colleagues. Nguyen and Dewachi would continue to collaborate on conflict and the wounds of war for years to come.

Ultimately, Dewachi was able to return to the United States, but not as a citizen of Iraq. The country that had bombed, invaded, and left his own country in isolation had rendered him a refugee, a category he reluctantly accepted because it was the only way to return to Harvard to finish his PhD. After being awarded his degree, Dewachi moved to the American University of Beirut and started the Conflict Medicine Program with Abu-Sittah. As he was Iraqi himself, he was able to build trust among Abu-Sittah's large Iraqi patient population. He quickly began amassing data.

What he and Abu-Sittah learned was that the American invasion had not only razed Iraq's health-care system and left behind the detritus of war, from bullets, shrapnel, and shells to contaminated soil and water, but it had indirectly affected the supply of drugs in the region. With few immediately available options, when Iraqis fell ill they turned to anyone who promised to supply them with drugs. Few drugs of any quality were available. Soon the pharmacists (some real and some not) found that the front-line antibiotics would not work, so they started giving injections to anyone who could afford them. Injections were promised to be fast acting—tablets were considered slow. Leading antibiotics like carbapenems flowed freely. But quality control was nonexistent, accurate diagnosis was rare, and medical oversight severely lacking. As a result, many patients ended up in hospitals in Jordan or Lebanon.

Dewachi stayed in touch with his friend Nguyen, and together with Abu-Sittah they started to investigate the link between the Iraq Wars and the rapid rise of *Acinetobacter baumanii*. There was no smoking gun, but a number of strong indicators emerged as a result of their collaboration. Since 1994, scientists had known that heavy metals can cause drug-resistant *Acinetobacter*.[13] While these metals were not endemic to the region, the modern weapons of war contain them. Then there is the collapse of infrastructure, from hospitals to sewage lines. The consequences can be doubly damning. Broken sewage lines can spill into water supplies, and cement and metals used in building materials, once blown to pieces, can likewise contaminate the water. The team concluded that this kind of contamination would lead to resistance.

But is war the cause of resistance? Or is there just a correlation between the two? No one can do a clean experiment here to find out. The answer is unclear and perhaps always will be. And the bacteria don't care. Causation or correlation, they are presented with circumstances and enough of them take advantage. They evolve, and evolve toward resistance. However material ascertaining blame and responsibility for the problem is (and for those suffering, having that answer would be profound), for the ever-increasing population of resistant bacteria, it is of no interest whatsoever.

Wars and infections have always accompanied each other. In

the twentieth century, infections among the wounded created a new challenge as drug-resistant infections became a serious issue for the patients and the army medics. These resistant infections were seen by Cutler in the battlefields of Europe during World War II, were investigated by Holmes during the Vietnam War, and had now appeared in their nastiest form during the Gulf War. Indeed, it is one of the points of war. Unquestionably, when invading Iraq, America, like invaders from time immemorial, intended to degrade the country's ability to resist. The bombs dropped, the weapons used, were for the purpose of inflicting harm and trauma. And during occupation and enforced international isolation, the goal, again, was to inflict on Iraq harms sufficient to cause it to behave in ways more aligned with America's interests.

But, once more, bacteria don't care. They obey no borders, harbor no national loyalty, and are always self-preserving, self-advancing, and self-replicating.

CHAPTER 23

COUNTING THE DEAD

Cooum and Adyar are short rivers by Indian standards. Consider the Ganges, which starts in the Himalayan Mountains, fed by its glaciers, and runs 1,569 miles across the country before spilling into the Bay of Bengal. The Cooum and Adyar barely make a total length of a hundred miles, though they trisect South India's cultural and economic center, Chennai—a city that was called Madras when Ramanan Laxminarayan was growing up there.[1]

Laxminarayan excelled throughout high school and then enrolled in an engineering program. Yet from the outset he had doubts about his chosen field. It wouldn't, he thought, present him with solutions to the problem that was bothering him. The two rivers that defined his city were massively polluted. As to the cause of the pollution, the powers that be had little doubt. The finger was always pointed at the people who lived downriver—close to the most polluted part of the city. These were also the poorest people. The ruling from upriver was that this population was entirely responsible for the very filth they lived in.

Laxminarayan wasn't convinced. He decided to look for himself, following the river upstream. What he found was quite clear: the trash and sewage from those living upstream would flow down, and because the water upstream enjoyed a more constant flow, it remained cleaner. Though the more well-to-do were the cause of the problem, the people who lived downstream lived with the consequences and were blamed for them too. This disparity taught him a valuable lesson. The problem of environmental pollution in Chennai was not an engineering problem; it required the tools of social science and advocacy.

From India, Laxminarayan went to Seattle, where he pursued his PhD studies at the University of Washington with Gardner Brown as his research supervisor. Brown combined two of Laxminarayan's interests—a love of the outdoors and viewing public health through the lens of social science (the two men always conducted their meetings on the hiking trails of Washington). And it was during this time that Laxminarayan became interested in antibiotics and antibiotic resistance. This was in the early 2000s, when only a small number of scientists and researchers were aware of the mounting problem of resistance. Among those few, Laxminarayan took a job at an organization called Resources for the Future in Washington, DC. And in 2005, two chance encounters changed his perspective on antibiotic resistance completely.

Laxminarayan was in a cab, heading to a meeting in DC one day. His driver was a young black man from Baltimore, and he asked Laxminarayan about his work. As Laxminarayan explained what he did for a living, and the purpose of the meeting he was going to, the driver had his own story to tell about drug-resistant infections. He had been in and out of the hospital as a result of resistant infection, in fact. The driver was being pushed deeper and deeper into poverty and despair because the doctors could not cure his infection. His persistent, resistant infection was making him desperate.

Around the same time, Laxminarayan met another man by the name of Michael Bennett, who introduced Laxminarayan, in a manner of speaking, to his father, Mark. The older Bennett was the son of Jewish immigrants from Ukraine and had grown up in the slums of New York City.[2] When World War II began, Mark Bennett served in the military and became a hero and veteran of the battles of the South Pacific. Not only did he survive the war, but he managed to defy the malaria he contracted while he was fighting there. He was a man who could not be broken by poverty, New York City, the casual anti-Semitism he encountered, or the ravages of war. But then something broke him completely: MRSA.

Repeated trips to the hospital due to various ailments exposed Mark to an infection that was slowly sucking the life out of him. The doctors who treated Mark were often careless, confused, or seemed indifferent to his suffering. Now, Mark's son Mike was doing all that he could so others who face the battle with drug resistance do not get the same treatment as his father did while he was in the hospital repeatedly for one infection or another. It was a story that Mike recounted to Laxminarayan during a hearing at the Capitol, and it resonated with the congressional staff whom Laxminarayan briefed. The problems afflicting the poor in his home city way back in India were afflicting humanity across the globe. Laxminarayan decided to take action, and his focus would be MRSA.

Laxminarayan first investigated the extent of the problem. This seemingly obvious question had never been addressed. No one had quantified how many people die annually from MRSA. In 2007, Laxminarayan and his colleagues published a study that showed that in a six-year period, from 1999 to 2005, MRSA-related hospitalization had increased by over 60 percent.[3] In 1999, the number of MRSA-related hospitalizations in the United States was less than 130,000. By 2005, it was close to 280,000. Backed by data, Laxminarayan argued that MRSA should be declared a national priority for disease control. Over the next decade, several studies led by Laxminarayan and colleagues would reshape the national, and ultimately the global, debate.

To systematically look at the issue of drug resistance, Laxminarayan created a new entity, called the Center for Disease

Dynamics Economics and Policy, with offices in Delhi and in DC. The center quickly became one of the global leaders in estimating and quantifying the burden of antibiotic resistance. If the facts could be baldly presented, perhaps those in power would point their fingers toward the actual culprits and act.

In 2016, Laxminarayan and his team published a highly provocative report. They found that the number of children under five who die every year around the world due to resistant infections was about 214,000.[4] This was largely due to exposure to bacterial infections that had become resistant. However, about twice as many died because they had no access to antibiotics whatsoever. The report exposed the tension already felt by others between access and excess.[5] Which was worse: using antibiotics without prescription in an incorrect way or not having any?

Laxminarayan explained that the issue of access and excess especially hurt those at the bottom of the economic pyramid. As resistance increases, the first-line drugs that are often cheap and highly subsidized for the poor communities become useless for drug-resistant infections. This means people who do not have access to the finances necessary to buy more expensive, second- and third-line antibiotics will always be worse off. They either cannot afford any antibiotics or can afford only those that have no efficacy against resistant infections—the antibiotics they can access work only when the bacteria are not resistant. After analyzing the burden of antimicrobial resistance (AMR) in the United States, Laxminarayan focused his work on India and China, countries he believes are going to shape the global landscape of antibiotics.

When Laxminarayan started his work on water pollution in the early 1990s, little was known about the burden of waterborne antibiotics. By the mid-2000s, this would change. Like the polluted rivers in South America that led to the typhoid outbreak in Aberdeen in 1964, Indian rivers continued to be vast reservoirs of pathogens. But things are much worse now. The polluted rivers now carried antibiotic-resistant bacteria. These rivers also contained both antibiotic residues coming from human and animal waste as well as antibiotics that were dumped in whole. The problem of high levels of antibiotics in water becomes unmanageable during floods—such as those experienced in Chennai in 2015.[6] Among the waterways that contain the highest levels of antibiotic resistance are the holy rivers of the Yamuna and Ganges, as well as the Cooum and Adyar. The problem is not just in India. As waterways become toxic with antibiotics, it affects everyone— even those who are not taking any drugs. And if antibiotics end up in water, scientists wondered if water might hold other clues about the global size of the problem and which communities might be at greatest risk. The answer brought them not to large rivers but to sewage.

CHAPTER 24

CLUES IN THE SEWAGE

Frank Møller Aarestrup faced a challenge.[1] A professor at the Technical University of Denmark, located about ten miles from Copenhagen, Aarestrup wanted to test the fecal matter of the world, on a budget. Then a lightbulb went off: Kastrup.

The Copenhagen airport has been in operation since 1926. It is the largest airport in the Nordic countries, a gateway to Scandinavia, and a hub for global air travel. Located in the small town of Kastrup, from which it takes its name, on the densely populated island of Amager, the airport is also very close to the city center.

In 2015, Aarestrup put his request to the airport's authorities. He wanted to collect human waste, liquid and solid, from long-haul flights from the United States and Asia. The airport had expanded significantly since 2000, and in the last couple of years, flights from Asia and the Middle East were bringing in hundreds of passengers a day. This was in addition to the direct flights on SAS airlines from the United States and Canada. The request was unusual enough that it gave the authorities pause. Why would anyone be interested in the human waste of fatigued passengers?

But Aarestrup had a reason, and one that would invariably connect antibiotic resistance to socioeconomics.

———

Aarestrup had grown up on a farm in rural Denmark. After high school, he decided to become a veterinarian—a profession that would come in handy on the farm. But by the time Frank enrolled at the university, the scientific world was abuzz over the Human Genome Project. Frank found the rapidly exploding field of genomics thrilling. After completing his undergraduate degree, he stuck around and got his PhD in veterinary microbiology. The link between veterinary microbiology and genomics was nascent in the mid-1990s, and there was a huge opportunity to make a mark. Aarestrup got to work immediately.

During the early 1990s, the issue of antibiotic resistance in animals was politically charged in many countries, including Denmark. In neighboring Norway, active conversations about antibiotics in salmon feed were being debated at the highest levels of government. In Denmark, there was little regulation of antibiotics going into animal feed—which Aarestrup remembers as being treated more or less like vitamins for growth promotion. There was also little surveillance of the livestock industry—no one knew who was doing what and with which antibiotics.

Aarestrup found the gap in knowledge both fascinating and troubling. Over the next few years, he developed a surveillance program monitoring antibiotics use within the livestock industry in Denmark that became a model eventually adopted by many organizations around the world, including the FDA and the US Department of Agriculture.

While he was working on the surveillance of animals, Aarestrup became obsessed with another problem that had been largely ignored: the global surveillance of antibiotic resistance was insufficient. The question about the global consumption of antibiotics was simple—but largely overlooked. While there were some estimates about how many drugs were used globally per year in animals and humans—an understanding of how resistance and consumption were connected was lacking. The policy makers and scientists also had little information on how consumption patterns were related to national policies governing drug use or socioeconomic factors. The only data coming to international agencies was submitted from doctors who were treating the patients. But this data was woefully incomplete, prone to the biases of the doctors, and focused on the rarer cases that rise to the attention of physicians to begin with. It also produced no information on various resistance genes that may be present in a particular population.

Aarestrup, now leading a team out of his laboratory, decided to do something about this gap. However, the team faced an immediate challenge. To get at the data, they had to get samples. And whether one was collecting cheek swabs or feces, doing so around the globe and in quantities sufficient for study seemed neither feasible nor ethical. Aarestrup's team wondered if there could be a way to analyze large samples that had already been pooled together. And one place accumulated those samples: toilets. Our fecal and liquid waste has bacterial genes that can be analyzed. The study had to be conducted in a systematic fashion. And samples had to come from around the world and be collected without breaking the bank.

If the waste from long flights could be collected and analyzed, the team might be able to see how the presence of resistance genes correlated with various regions of the world. Once the authorities were convinced that the request was not a prank, the team collected waste from eighteen international flights arriving from nine cities in three different parts of the world: Bangkok, Beijing, Islamabad, Newark, Kangerlussuaq, Singapore, Tokyo, Toronto, and Washington, DC.[2] Each flight contained approximately 400 liters of human waste. Good enough. Frank and his team did not need tons of waste—just representative samples. Once collected, the samples were analyzed in Aarestrup's lab using whole genome sequencing, a process that enables researchers to read the entire DNA blueprint of an organism. In this case—it was going to be about the bacteria in the sewage samples, and it would yield definitive information on whether there were resistant mutations present in these bugs.

A pattern of global resistance began to emerge. The analysis showed that flights from South Asia had a much higher abundance of genes with resistance to Beta-lactams—the class of drugs that contains penicillins, carbapenems, and cephalosporins.[3] *Salmonella enterica* and norovirus were also more prevalent in South Asia than in North America. And one takeaway from the study was immediate: a policy that might work very well in the United States might not work in India.

There were also some problems: the team was sampling the waste from people who were finishing their journey on the plane heading to Copenhagen. But they could have started that journey from almost anywhere, via a connecting flight. The data collected couldn't point to problems in their hometowns. Nothing could be said conclusively about their socioeconomic status either.

Aarestrup knew that there was only one solution to these issues. The team could not restrict itself to looking only at waste arriving in Copenhagen—it needed to look at untreated domestic sewage in the home countries of the departing planes as well. The samples had to be collected not at Kastrup but in untreated sewage plants in cities on six continents. The study was about to increase in scope in dramatic ways—and it was going to be expensive.

Aarestrup started to crowdsource the project. He wrote to his colleagues, collaborators, and friends, asking if they would be interested in getting involved. His team created a mechanism for teams all over the world to indicate their willingness to participate. Eventually, they had sufficient buy-in, and the team launched the Global Sewage Surveillance Project,[4] which spelled out a long and detailed protocol. Samples were collected from seventy-nine different sites in sixty countries, and, following protocol, everything was collected on two consecutive days between January 25 and February 5, 2016.[5]

From Nigeria to Nepal, Peru to Pakistan, Togo to Turkey—samples were collected, packed, and sent to Aarestrup's lab at the Danish Technical University. There, a team of students and researchers analyzed the pretreated waste and checked genes for resistance.

As intended, they now had a vast quantity of data that yielded the expanded study's first major insight. Antibiotic use was not the dominant factor in the global AMR burden. Instead, a much deeper problem surfaced, and that was poverty.

The populations of richer countries in North America, Europe, and Australia had fewer genes associated with resistant bacteria, whereas those of poorer countries in South Asia, Latin America, and sub-Saharan Africa had far more resistance-conferring genes. Countries such as India, Vietnam, and Brazil, where poor sanitation and nutrition were prevalent, were producing far more genes that are responsible for resistance. In countries such as Sweden and New Zealand, where rules and regulations governing hygiene were strongly enforced, the burden of AMR genes was significantly lower.

The study's results were enlightening—and sobering, because despite hundreds of billions of dollars already spent on trying to address the problem, it was still very much a global reality. And the sewage and hygiene problem was about to put a city in the south of Pakistan on the global health map. The reason? It would soon become the epicenter of the first extensively drug-resistant typhoid—the same disease that had caused widespread panic in Aberdeen in 1964, but this time it was much worse, and the arsenal of available drugs to treat it had shrunk dramatically.

CHAPTER 25

X IS FOR EXTENSIVE

Rumina Hasan wasn't expecting anything unusual during her rotation on the blood culture bench in December 2016.[1] An experienced clinical microbiologist and pathologist, she oversaw dozens of staff members and junior doctors in her department at Aga Khan University Hospital (AKUH), Pakistan's most renowned. Named after the spiritual leader of the Ismaili sect of Shia Islam, the hospital sees patients not only at its sprawling campus in Karachi, but also at the hundreds of branches across the country that provide laboratory and diagnostic testing services.

December in Karachi is fairly pleasant, with temperatures between the low fifties and high seventies. Hasan was busy at the time, going through blood culture samples, when something caught her eye. The results did not quite make sense. She had seen her share of unusual blood cultures, but this was concerning. She reran the tests, which confirmed her finding. The sample in front of her was resistant to ceftriaxone, a drug used routinely for treating typhoid.

Drug-resistant typhoid is not unusual in Pakistan and in many other parts of the world where sanitation problems are acute.[2] While the outbreak in Aberdeen had caused panic, there were a whole host of drugs that were available to treat patients at the time. Since the 1970s, however, typhoid has become gradually resistant to an arsenal of drugs.[3] Reports from Mexico in 1972 showed that the frontline drug against typhoid, chloramphenicol, was no longer effective. In the 1990s, the next line of drugs, ampicillin, amoxicillin, and the sulfamethoxazole-trimethoprim combination, were proving to be ineffective. Clinicians switched again, this time to the class of drugs called fluoroquinolones. But by the early 2000s, this set of drugs was also becoming ineffective. The most potent drug that doctors had left was ceftriaxone.

Initially, only one sample exhibited signs of resistance to the drug, but as Hasan started to pay close attention to the issue, she noticed a trend. Over the coming days, as Hasan and her colleague Dr. Sadia Shakoor tested additional samples, an increasing number showed ceftriaxone-resistant typhoid. Hasan and Shakoor investigated further. They found that all of the resistant samples were coming from a single city—Hyderabad, about 100 miles northeast of Karachi. Hasan and Shakoor called colleagues there and asked them to investigate as well. They called up pediatricians in charge of a children's ward in Hyderabad. The Aga Khan team, which now included Dr. Farah Qamar, a pediatric infectious-disease specialist, reached out to colleagues at the Gates Foundation, who started sampling water and sewage in Hyderabad and initiated efforts to vaccinate children against typhoid in the city. Hasan also contacted the provincial authorities, who didn't seem too concerned.

More had to be done. Hasan and her team started sending weekly reports to the local government in Karachi and then to the Pakistan National Institutes of Health in the capital, Islamabad. Hasan wasn't willing to give up, even in the face of apathy from the government and lackluster interest from the media.

All her life, from her time studying at a medical school in England to her early professional years in Pakistan, Hasan had learned to push boundaries. From working in the leprosy ward in Karachi to being one of the founders of the Pakistan Antimicrobial Resistance Network (PARN), she knew how to persist. So when the response from the Sindh authorities was lukewarm, with local and governmental powers expressing little interest, or even capacity, to hear the scientific basis of what was going on, she didn't shy away or back down. She started to look farther afield. The facilities at the national institute in Islamabad were far from cutting edge. In addition, the layers of bureaucracy clogged up the decision-making process in Pakistan. If Hasan was going to map genetic markers for resistance in order to understand whether this was truly a new scenario of multi-drug-resistant typhoid, she would need more help.

Luckily, one of her junior colleagues, Dr. Zahra Hasan, had worked with Professor Gordon Dougan, an expert on the genetic markers of resistance in infectious diseases at the Sanger Institute in Cambridge, England. Zahra reached out to Dougan. Initially, there was not much interest coming from his group; they receive requests for help with genetic analysis of pathogens on a regular basis. But Zahra persisted, and Dougan finally agreed to look at a selection of her samples. Zahra's team chose to send a total of one hundred samples, eighty-nine of the resistant ones and eleven that were sensitive to the frontline treatment.

Around the same time, a new postdoctoral fellow in Gordon's lab connected with Rumina Hasan and her team in Karachi. Elizabeth Klemm had recently moved to England after receiving her PhD from the Massachusetts Institute of Technology. She became the point person in the collaboration between Dougan's lab at the Sanger Institute and Hasan's group at Aga Khan. She quickly understood that the frontline therapy for typhoid wasn't working and the hospital in Pakistan was running out of options. Klemm, with Dougan's permission, moved the eighty-nine samples from Pakistan to the front of the queue.

Klemm was familiar with recently published studies pinpointing the drug-resistant typhoid in Iraq, Palestine, Pakistan, India, and Bangladesh. But never before had there been an outbreak like the one Rumina was reporting. Once she looked at the results, Klemm knew why the strain of typhoid was resistant to all of the antibiotics. The strain carried genes that conferred resistance to chloramphenicol, amoxicillin, ampicillin, and TMP-SMZ. This wasn't all—the resistant strain was also carrying a mutation that enabled it to become resistant to ciprofloxacin. This finding was consistent with what Hasan's team saw. What was alarming was this bacteria for typhoid, *Salmonella typhi*, had picked up a plasmid—the mobile unit of DNA that Lederberg and Watanabe had shown as a cause for multiple-drug resistance—from *E. coli*. This new mobile DNA had made the typhoid-causing bacteria resistant to one more class of drugs, ceftriaxone.

The team compiled the findings in a paper that they submitted to the prestigious journal, *The Lancet*. The editorial board moved slowly, requesting more clinical data. But Hasan, now with Klemm's avid support, persisted yet again. Finally, their work was published in January 2018.[4] Soon the news was making headlines around the world.[5]

This time, the Pakistani government paid attention, but its options were limited. The hope for most of the patients was a single drug, azithromycin. Carbapenems were another option, but an expensive one that was beyond the capacity of a poorer public health system. The carbapenem of choice also requires an intravenous infusion, which requires sanitation standards that are difficult to achieve in rural hospitals.

By December 2018, nearly five thousand people in Pakistan had been afflicted by this resistant strain of typhoid.[6] It was the first known case of an extensively drug resistant (XDR) outbreak of typhoid. The CDC issued a warning for people traveling to Pakistan and reported that some of the patients seen in the United States with extensively drug-resistant typhoid had recently been in Pakistan. For now, every international agency was recommending using azithromycin, but researchers knew it was only a matter of time before the last line of defense would start to crack.

CHAPTER 26

TOO MUCH OR TOO LITTLE?

From 2014 to 2017, azithromycin was given—by an international team of researchers and clinicians from the United States and Europe—as a prophylactic, a preventative measure, to tens of thousands of children in a large clinical trial undertaken in Niger, Tanzania, and Malawi.[1] All of the children were under five years of age and were given the drug whether they were sick or not. Over the course of the study, they received the prophylactic doses every six months, for a period of two years. The children were divided into two nearly equal groups, and about ninety-seven thousand were given the drug, and the ninety-three thousand in the control group did not receive the drug and got only the placebo.

The results were remarkable.[2] They were most significant in Niger, the poorest of the three countries in the study. The mortality rate was 18 percent lower in the group of children in Niger who got azithromycin than those who did not. In Malawi and Tanzania the numbers were much smaller and statistically less significant. Within the group of children whose life mortality rate had declined by 18 percent, the biggest impact was among the youngest children, who were under six months old. Here the improvement was close to 25 percent.

In the global effort to improve child survival, numbers like these are rare. This was big news in 2018. The conclusion was clear: prophylactic azithromycin could save the lives of vulnerable children in poor countries. And when the report came out, there was excitement—and outrage over the prophylactic use of a potent antibiotic, a global resistance to which could devastate communities where this drug was a last hope and lifeline.[3]

The scientist leading the study was Thomas Lietman, a professor at the University of California in San Francisco.[4] Educated at Yale and trained in ophthalmology at Johns Hopkins, Lietman arrived at UCSF in the late 1990s. An eye specialist may not come to mind when we think about antibiotics research, but the link between the two goes back decades.

The connection is trachoma, an infective eye disease resulting in blindness if left untreated. Caused by the bacterial infection *Chlamydia trachomatis*, the affliction has been around since the Bronze Age. And in 1897, it was the first disease classified by the United States as dangerous and contagious.[5] Immigrants trying to enter the United States had to be tested for trachoma and were sent back to Europe if they had the disease. President Woodrow Wilson signed an act in June 1913 to release funding to eradicate trachoma, and with improvements in hygiene, increased awareness, and better treatment options, it was, in fact, eradicated in the country. But in other parts of the world, particularly Ethiopia and South Sudan, the problem persisted.

In the late 1990s, scientists discovered that antibiotics, particularly azithromycin, can cure trachoma.[6] Then in 2008, a limited trial in Ethiopia showed that mass administration of azithromycin could improve trachoma treatment and prevent the spread of the infection substantially.[7] But there was an unexpected result. Mass administration of azithromycin as a prophylactic also appeared to decrease child mortality rates across the board.

Because the 2008 trial had not been designed to study mortality, its conclusions were not definitive. But many scientists, including Lietman, were intrigued and decided to investigate further and push their study beyond trachoma. Could giving azithromycin really improve the chances that a child could survive the harsh realities of life in many parts of Africa?

They approached funding agencies and proposed a very large clinical trial, where they would give azithromycin to children under five and also do a control with placebo. Nothing of this sort had been done in recent history. They chose Niger, Malawi, and Tanzania. The Bill & Melinda Gates Foundation was one of the funding agencies that Lietman contacted, and after much back and forth, they agreed to help.

Lietman put together a consortium of scientists from multiple institutions to conduct the trial. As is often the case, the project progressed slowly. It took nearly three years for the necessary approvals to come through. But once they did, the scientists were there, at the ready, eager to rigorously test their hypothesis that prophylactic mass administration of azithromycin can prevent infant mortality.

The study was called MORDOR, an acronym taken from the French (the official language of Niger), Macrolides Oraux pour Réduire les Décès avec un Oeil sur la Résistance or Mortality Reduction Through Oral Azithromycin. Given the wedge it would soon drive in the scientific community, that the catchy name also invoked the mythical land of Mount Doom in Tolkien's *Lord of the Rings* was fitting.[8]

By 2018, the results were clear. Two doses a year, for two years, could significantly improve a baby's chances of survival in Niger. The Gates Foundation celebrated the fact that there was an intervention within our reach that could reduce infant mortality by significant numbers. But not everyone was thrilled.

There were unanswered questions related to the reasons why prophylactic azithromycin worked so well. The Lietman team didn't have a clear answer. They had a number of hypotheses—perhaps the drug was helping the kids survive malaria, or perhaps it was changing the microbiome of the children, or perhaps it was helping them fight diarrheal or respiratory infections. All were plausible; none were proved. And the answer "we do not know for sure" was not comforting for scientists and clinicians who wanted to know why a single drug, given just twice a year, could do something that other far more complicated interventions couldn't.

But there was an even bigger elephant in the room. In this day and age of mounting bacterial resistance, how could anyone justify giving antibiotics without any cause? Weren't the Gates Foundation and Lietman playing with fire? In many countries, such as Pakistan, azithromycin was a last-resort drug. And here was a team of scientists and specialists freely experimenting with it and giving it to children who weren't even sick.

In addition, if Niger could make this a policy, would other countries follow suit? Clinical trials are very expensive, and there is no way countries across the world could conduct the same careful trial that had been conducted in Niger. But without such a study, and without anyone knowing why the drug had produced the results it had, how should each country make a decision about conducting its own mass administration of azithromycin? If it is acceptable for Niger to give drugs to its children, why is it not acceptable in other countries?

Yet another familiar question raised its head. Public health researchers who work on the consequences of counterfeit and substandard drugs were also alarmed. Niger and other developing countries have almost no regulations in place to enforce quality control. Wouldn't there be an easy opportunity for peddlers of fake drugs to enter any country considering the mass use of azithromycin? Once drugs of compromised quality are available, it's very difficult to fix the consequences, namely increased resistance. And the mass spread of azithromycin into the environment was also a concern. As we know, antibiotics pass through our systems and end up in the environment as waste in water and soil. In countries with poor sanitation, this would mean yet more people and animals would be getting the drug.

Lietman was aware of all of this—and did not dismiss the criticism. He agreed that if the mass prophylactic use of antibiotics is made a policy, it would select for resistance. But he asked the hard question: How can we allow 10 percent of our children to die before their fifth birthday? We have tried to improve water and sanitation for decades, perhaps fifty years, and yet we haven't made any real progress. If we know that there is a simple intervention that saves the lives of babies and kids, why would we not do it?

Other scientists posed another question: Aren't we presenting parents of a young baby with two choices? We can decide not to intervene now and let the children have a high chance of mortality, or reduce the chance of mortality now at the likely cost of increased resistance twenty years later. Which would they choose? When I put the question to Lietman, he responded in turn with a question: "If we were to ask this question here in the United States, what would we do?"

Over dinner that night, I asked my wife what would we do if we had to make a decision about prophylactic use of antibiotic, or a high risk of child mortality. Our initial reaction was the same: giving prophylactic antibiotics to everyone, whether or not they were sick, was wrong. Being from Pakistan, where azithromycin is the last hope of many suffering from extensively drug-resistant typhoid, we knew how important it was to preserve its efficacy. Friends and family in the last year alone have had to rely on azithromycin as they battled typhoid. But as we deliberated, and looked at our two kids, and imagined ourselves in Niger, we couldn't say with certainty that we would decline the treatment.

CHAPTER 27

VISA NOT REQUIRED

"AMR does not need visas"—announced the head of the World Health Organization, Tedros Adhanom Ghebreyesus, at a World Health Summit meeting in Berlin in 2017. What he meant was that it is impossible to quarantine resistant pathogens. No walls, no barriers can stop them. The carriers are not just humans in planes, but the original inspiration for the modern flying machines: birds.

Drive seventy-five miles north from Dhaka, the capital of Bangladesh, and you can visit the exotic-bird market in Mymensingh. It's a rare and magnificent tapestry of colors, sights, and sounds. There, sellers collect a vast assortment of birds, not just those taken from local forests but birds that have been shipped to the market from far-off tropical forests. And there, affluent parents and their children pay for the expensive birds with their brilliant plumage and take them home as pets.

Dr. Md. Tanvir Rahman is a veterinary microbiologist with a long-standing interest in antimicrobial resistance.[1] While there is no national data on the extent of antibiotic resistance in exotic birds in Bangladesh, Tanvir knows from his own experience that the problem is widespread. He is well read and knows that once the problem expands, it will be nearly impossible for any animal to be safe.

Passing through Mymensingh's exotic-bird market one day,

Rahman wondered if even these exotic birds were also carrying resistance genes. Could the wild birds be just as vulnerable as the ones on the poultry farms? Rahman searched the research literature, spoke to veterinarians and microbiologists, and soon confirmed that no one had collected any satisfactory data such that he could answer his question. He decided to do the work himself.

Rahman did just what Stuart Levy had done four decades ago in using animal droppings to study the burden of antibiotic resistance in animals. He collected the droppings of birds from the pet shops and started analyzing the samples. In addition, he focused on migratory birds, not on the ones that were raised in captivity. He was wondering whether birds in the wild were also carriers.

As his results came in, his worst fears were confirmed. The bird droppings had bacteria that were resistant to a whole host of antibiotics, including colistin, chloramphenicol, erythromycin, ertapenem, azithromycin, and oxycyclin. Similar to Aarestrup's studies, the discovery raised unanswered, and perhaps unanswerable, questions. Rahman does not know how these birds picked up the resistant bacteria. They could have been given feed laced with antibiotics after being captured, or they could have been exposed to the bacteria in the environment. What is certain, however, is that these birds were carriers of resistant bacteria. These exotic birds were bringing more than just their colors and songs to new locations. The problem was no longer in confined pig farms in China, buffalo farms in Pakistan, or chicken farms in India. The bacteria were finding all possible ways to travel—sometimes in the bellies of birds that moved across countries and continents. Solutions would have to be transnational and transcontinental. And Scandinavia, a pioneer in animal vaccination and resistance surveillance through the work of scientists like Tore Midtvedt, offered yet another global solution.

In September 1961, the newly independent Democratic Republic of Congo was ripping itself apart at the seams. Civil war seemed inevitable. Separatist movements were declaring parts of the country independent. In January of that year the Congo's old colonial masters, the Belgians, along with America's Central Intelligence Agency, had executed the country's first, and dynamic, firebrand prime minister, Patrice Lumumba. The looming crisis was getting worse by the day.

In response, a UN flight was dispatched to southern Congo, its delegation hoping to broker a peace deal. There were sixteen people on the flight. And then, nine miles from Ndola, which was part of Northern Rhodesia (modern-day Zambia), the plane mysteriously crashed. Traveling with the delegation was the UN secretary general, a Swedish man named Dag Hammarskjöld. He was a consummate diplomat, one whom President Kennedy had referred to as the greatest statesman of our century. Already in the midst of a civil war, in the months after Hammarskjöld's death, Congo descended further into years of bloodshed that took an estimated one hundred thousand lives.[2]

In 1962, the Swedish government honored the famous diplomat by creating the Dag Hammarskjöld Foundation in Uppsala, the same town where he is buried. The foundation seeks to tackle the pressing problems of our time through what it calls a "meeting of the minds." Its guiding hope is that diplomacy can untangle the gravest threats to humanity.

Forty years after Hammarskjöld's death the foundation received an unusual proposal from Otto Cars, a professor at Uppsala University.[3] He wished to convene an international meeting on a subject that was indeed a global threat, but one that was on the margins of what the foundation was used to funding. There was no conflict, no warring parties. The problem, nonetheless, was international and directly connected to preserving precious, life-saving resources. The foundation was unsure, questioning whether the goals of Cars's proposal fit its agenda, but eventually agreed to fund the conference.

The result was ReAct, one of the most prominent networks of scientists, clinicians, public health activists, and policy makers.[4] The meeting that the foundation funded in 2005 was attended by sixty participants from twenty-five countries. Today ReAct is global, with a presence on five continents—Asia, Latin America, Africa, Europe, and North America. Its home base remains at Uppsala University. And its mission has grown with its reach. ReAct aims to coordinate globally what others had tried to implement locally, or even regionally—increase surveillance, generate awareness, prevent infection, and above all create a path for effective global policy.

Cars has since become a familiar face the world over, rallying support for research into antimicrobial resistance. It has been a lifelong interest of his. The first time he saw the extraordinary effectiveness of the drugs was in the late 1950s when his sister developed an infection in her eardrum and had to take penicillin. The same sister, a couple of years later, contracted scarlet fever and had to be isolated and quarantined. Antibiotics had to be used again, and their worth as a lifesaving precious resource was obvious.

Cars became an intern in the hospital where his mother was a nurse. In truth, he was more of a helping hand than an intern— and he helped with anything and everything that benefited from the almost limitless energy of a teenage boy. When the time came for Cars to decide what he wanted to do with his life, he narrowed it down to medicine or law. Cars chose medicine, and he made infectious diseases his specialty. Sweden had long elevated the danger of infectious diseases in a way that America hadn't.

One of the hospitals that was focusing on infectious disease was in Uppsala, which is where Cars started his career. Throughout their training, Cars and his fellow Swedish doctors were instructed to appreciate these drugs, and they were trained to prescribe only when absolutely necessary. It was an understanding that Cars took with him into his own research, which started with investigations into the proper dosing and concentrations for antibiotics. But despite the governing protocols and the widespread appreciation of the need to treat antibiotics with respect, there was a gradual increase in their usage not just in Sweden but in the Nordic countries in general.

In the early 1990s, there was an outbreak of penicillin-resistant pneumococci among children in the southern part of Sweden. That it could happen in Sweden bothered Cars deeply, and he also knew that more outbreaks were just around the corner if prescription and sales patterns continued to increase. Armed with a sense of urgency and general apathy about changing the status quo, much like Stuart Levy through APUA in the early 1980s, he decided to act at a global level.

In 1995, Cars and his colleagues started to track the sale of antibiotics in a program called the Swedish Strategic Program Against Antibiotic Resistance—known by its Swedish acronym STRAMA. Clinics and hospitals, much to Cars's surprise, participated willingly, sharing voluntarily information on the number of antibiotics their doctors were prescribing. Thereafter, they also underwent efforts to reduce the amount. The program was a huge success. The Swedish public health department paid attention and eventually brought STRAMA under the umbrella of its national program. Over the years, ever more health organizations and physicians started participating, sharing their information and strategies on how to reduce prescriptions and sales.[5] Twenty years later, similar programs have been introduced in multiple countries.

And Cars wasn't done. While STRAMA became a global model for change, he knew that Sweden was not representative of all countries. If resistance was going to be meaningfully addressed, he needed to engage professionals from all over the world. In

2001, the World Health Organization planned to send out a global call for action, and the stage was set for the big launch in Washington, DC. The date was September 11, 2001.

The global events of that day forced WHO to postpone the meeting. As the political situation stabilized, and Sweden pushed for a renewed effort, the interest at WHO started to wane. Data supporting the extent of the problem was lacking. While Sweden was prepared to advocate for global awareness, it seemed the world was not prepared to listen. And there was an additional problem. Antibiotic resistance did not fit the existing framework of global responses to disease. There was no straight answer to the question people would ask about which disease this awareness campaign was about. They expected to hear an answer like typhoid or cholera—but there was no such simple answer. So Cars decided to approach the Hammarskjöld Foundation.

In 2009, ReAct got the opportunity to shape the global agenda. That year the rotating presidency of the European Union would fall to Sweden, which meant that the EU summit would be held in Stockholm. With all of Europe coming to their backyard, Cars and ReAct were invited to attend and help shape the union's health agenda. For the first time at such a meeting, the issue of antibiotic resistance would take center stage. And this in turn led to the EU adopting a comprehensive action plan.

With ever greater urgency, the world was recognizing and starting to respond to the global threat. The WHO Global Action Plan (GAP) subsequently launched in 2015 would incorporate the various pieces of advocacy developed by ReAct, STRAMA, and other institutions to create a platform for countries to adopt. The plan would have specific targets and goals that countries could modify and adapt to their unique needs. For some countries, it would mean increasing awareness; for others, it would mean investing in infrastructure to conduct better surveillance. It was time for everyone to own the problem. The hope was that if every country contributed, the world could turn the tide against antibiotic resistance.

The precious resource—which had held Cars's attention as a boy, galvanized his energies as a scientist, and led him to the launch of an organization that has changed the course of humankind's approach to protecting that resource—stands as a testament to his life's work. Looking back on a mission that has consumed his life for the last four decades, one issue still worries him. What if no one bothers to make the resource in the first place? Because right now, pharmaceutical companies are jumping off the antibiotic ship in droves.

CHAPTER 28

THE DRY PIPELINE

The headline in *Bloomberg News* on July 11, 2018, read "Novartis Exits Antibiotics Research, Cuts 140 Jobs in Bay Area." The news came as a bombshell to researchers and public health professionals. Not only was this yet another case of a large pharmaceutical company distancing itself from the antibiotic market, but Novartis had more than thirty potential drugs in the pipeline. When the Swiss company presented its reasons to the public, it used familiar language. The companies that had already left the market had all said something similar. The spokesperson for Novartis said the company wanted to "prioritize our resources in other areas where we believe we are better positioned to develop innovative medicines."

Two months before the Novartis announcement, Allergan, an Irish company, decided to divest its $1.5 billion from infectious disease into other areas such as eye care and diseases of the nervous system.[1] Two years before that, AstraZeneca, another global giant in pharmaceuticals, had sold its antibiotics business.[2] Out of all the major pharmaceutical companies, only four (Pfizer, Merck, Roche, and GlaxoSmithKline) maintain an interest in antibiotics. And given the realities, it's fair to wonder if they're also eyeing the exit.

Pharma's skepticism toward antibiotic research is based on history and financial realities. The decades of the 1950s and '60s were a honeymoon period, during which everyone wanted to get a piece of what looked to be an ever-increasing pie. But it ended, abruptly. The last discovery of a new class of drugs for Gram-negative resistant infection was nearly six decades ago, in 1962. That was nalidixic acid, the precursor to quinolones and fluoroquinolones. And since 1984, no new class of drugs has come to the market.[3] Since then all new drugs are, in fact, modifications of previously existing classes. Even as the pipeline of drugs has dried up, development costs have sharply increased. The development costs rose from $231 million in 1987 to $802 million in 2001.[4] The cost of clinical trials has also gone up, as regulators demand larger multi-country clinical trials and enforce more rigid guidelines.

Most antibiotics in the development pipeline haven't even made it to the market. Eighty percent of cancer drugs in clinical trials reach consumers, while only 2 percent of antibiotics get through. The cumulative consequence is evident. Right now, about fifty antibiotic drugs are in development. By comparison, in 2014 alone there were eight hundred different oncology drugs in clinical trials.[5]

The finances also look grim. Companies often measure their investments through a metric called net present value, or NPV. It's the difference between the present value of cash inflow and cash outflow. For oncology, the NPV is a net positive of $300 million; for neurologic disorders it is $720 million; and for musculoskeletal disorders like arthritis it's a whopping $1.1 billion. For antibiotics, the number is *negative* $50 million. Companies investing in antibiotics are likely to lose money.[6]

Then there's the final catch. After clearing all the hurdles—financial and regulatory—given all that we now know about resistance, a new antibiotic would be reserved for exceptional circumstances and used sparingly under strict supervision. For everyone focused on preserving a precious resource, that sounds marvelous. For companies interested in selling a drug to get a return on its investment, it sounds absurd. For profit-driven Big Pharma, getting out of the business of antibiotic development isn't a difficult decision.

Case in point, in the summer of 2018, the FDA approved a drug called plazomicin for treating complicated urinary tract infections caused by multi-drug-resistant bacteria, including those resistant to carbapenems. This was a major victory for Achaogen, the drug company based out of southern San Francisco—except that in the first year after FDA approval, sales of the drug were not particularly impressive. The company brought in less than $1 million. By April 2019, the company filed for bankruptcy.[7]

Pharmaceutical companies can no longer make the case that antibiotics are a good investment. But what if we're thinking about it all wrong? What if we outsource discovery to startup companies—not just in Boston or Barcelona, but in Beijing and Bangalore?

Perhaps there's a need to rethink the whole model of antibiotic discovery. Kevin Outterson, a law professor, aimed to do just that. What incubated his idea was access to medicines crystalized during one of the biggest public health crises of the last half century—the HIV epidemic in sub-Saharan Africa.

CHAPTER 29

NEW WAYS TO DO OLD BUSINESS

It had been six years since the apartheid regime fell in South Africa. Nelson Mandela was no longer the leader of the nation. But for many black South Africans, the horrors of apartheid were being replaced by another threat: AIDS. Thabo Mbeki, the president of the republic, was focused on the pressing concerns of diplomacy throughout the African continent and on jobs for the middle class. But sooner or later, it was assumed that he would have to confront the AIDS crisis that was destroying families and communities.

In July 2000, Mbeki stood at the podium to open the XIII International AIDS Conference, but instead of talking about HIV, he cited poverty as the biggest problem facing South Africa. He went on to discuss poverty-related diseases in general and, much to the anger of the researchers in the room, refused to say that HIV causes AIDS. Two months later, in the national Parliament, Mbeki said: "How does a virus cause a syndrome? It can't."[1]

Given Mbeki's publicly stated beliefs, his subsequent policies were ineffective and misdirected. They would result in the deaths of millions. But Mbeki was not the only challenge facing people who were living with HIV in South Africa. There was another problem: drugs were out of reach.[2] By 2000, the cost of effective drug treatments for HIV in South Africa was nearly $15,000 per year per patient—far beyond the means of those suffering from the virus. Many of them earned less than that figure in an entire year. In Western Europe, where most of these drugs were manufactured, the costs would be completely covered or heavily subsidized by the national insurance plans. Nothing of the sort existed in South Africa.

Out of the consequent, rising frustration, concerned South Africans drew upon the principles they had perfected during apartheid: citizen activism, awareness, and direct action. Zackie Achmat, along with ten other activists, founded the Treatment Action Campaign—an activism platform to protest the prohibitive costs of effective drugs to treat HIV.[3] At that time, the patents for anti-HIV drugs were controlled by three large pharmaceutical companies, which meant that they controlled the prices of the drugs and, as a result, access to them. But Cipla, a company based in Mumbai that specialized in manufacturing generics, entered the fold and said it would gladly make the drugs and offer them to South Africa at a fraction of the current price. Cipla didn't have the patent to make the drugs, but TAC didn't care about patents—they wanted the drugs for those who needed them, and they were willing to take the risk and work with the company despite the likely international pressure from big-name pharmaceutical companies.[4]

Several large pharmaceutical companies making the drugs dug in their heels. The CEO of the British pharmaceutical company Glaxo Smith Kline referred to the leadership at Cipla as "pirates" for stealing their intellectual property, as well as the income from the sales of their drugs.[5] As the contest unfolded, the South African government wavered as to whose side it would support. The local and international public pressure forced their hand, and they eventually sided with Cipla, agreeing to allow it to market the drug in South Africa. The large companies inevitably sued the government. They argued that it was in breach of contract and that it was actively working to undermine the interests of pharmaceutical companies by allowing generic companies (which did not have the patents or the licenses) to sell the drugs.

The government, burdened by bureaucracy and indecision, had a wavering position on making treatment affordable. The frustration among patients was growing. TAC, in turn, responded with a campaign designed to raise awareness of the HIV crisis and the lack of affordable access to treatment. It partnered with organizations around the globe and orchestrated demonstrations from London to New York, Mumbai to Melbourne. Supporters responded to what seemed a clear argument for access to drugs to preserve lives immediately at risk. The international pressure worked, and the pharmaceutical companies withdrew their lawsuit in a major victory for TAC, the AIDS patients, and the generic-drug companies.[6]

By 2005, drug patents and access to drugs were hot topics of debate throughout the global legal community. Some scholars sided with the drug companies: How could they maintain their ability to create new drugs and make new investments if they couldn't protect their patents in international markets? Others argued on behalf of the patients. Companies can protect their patents, but what good is a new drug if people can't afford it? With so many lives at risk, shouldn't organizations like TAC be applauded for their efforts? The global HIV crisis had opened up a battle over access, innovation, and pricing, all of which would take on special meaning in the context of antibiotic resistance.

A paper in a 2005 issue of the *Yale Journal of Health Policy, Law, and Ethics* obliquely raised the problem.[7] Its author, Kevin Outterson, then an associate professor of law at West Virginia University, laid out an argument for how pharmaceutical companies might balance access (which meant decreasing their prices) with their need to recover the costs of research. Deep in one of the footnotes of the one-hundred-page report, there's a statement that concerns the value of knowledge over time, and it speaks to the need for a new model in thinking about pharmaceutical innovation in the era of increasing drug resistance. "While knowledge is not destroyed through use," Outterson writes, "it may lose value because it is inappropriable."[8]

In brief, he was noting that an inventor cannot economically gain from her or his invention without help from intellectual property law, which is why typical patents are granted for twenty years. During that time, the inventor (or, more broadly, the holder of the patent) can exclusively sell the product. The reasoning is clear enough: the original inventor gets to have exclusive rights to profits for two decades, and after the patent expires, other companies can enter the market.

But there's an assumption in the way the law is written, namely, that after twenty years the product will still be beneficial. What if it's not? What if, after two decades, a drug loses its potency—or, worse, is harmful to the patient? The question haunted Outterson, for an obvious example was staring everyone in the face: antibiotics.

Antibiotics upended a core assumption about how intellectual property rights supported innovation. Once he identified the problem, Outterson became consumed with educating himself about the issue.[9] He studied and went to conferences, seminars, public hearings, and testimonies. He read up on the pharmaceutical companies and talked to their executives. He got to know the public health stakeholders, infectious disease physicians, and the members of the CDC.

Meanwhile, there were hardly any positive financial reasons for large pharmaceutical companies to continue making antibiotics. Resistance was progressing but, ironically, not fast enough to support sufficient innovation. Big companies increasingly abandoned the arena, focusing on profits related to cancer and diabetes drugs instead. Acknowledging the problem, senior officials in the FDA pronounced the antibiotic pipeline as "fragile."

Over the next ten years, Outterson continued to study the problem and got engaged in international discussions on how to increase investment in the discovery of new drugs. Globally, the world was also taking notice. The United States had lagged behind its European partners but that was starting to change. In September 2014, President Barack Obama issued an executive order regarding antibiotic resistance.[10] The order asked his council on science and technology to create a task force that would submit recommendations for a national action plan. He gave them six months. The action plan would be a blueprint for how the United States would respond to the problem of antibiotic resistance.

The task force released the plan on March 27, 2015, and it included all the right buzz words, scary statistics about the size of the problem, selective pressure on the bacteria to develop resistance, and stewardship to manage access and use of important antibiotics. But it also had something else. It had real dollars behind the words. The press release declared, "The proposed activities are consistent with investments in the President's FY 2016 Budget, which nearly doubles the amount of federal funding for combating and preventing antibiotic resistance to more than $1.2 billion."

The plan highlighted the need for a more robust pipeline of drugs, but large pharmaceutical companies were leery. For it to be worth their while, things had to be done differently. The task to do that fell on the shoulders of the Biomedical Advanced Research and Development Authority, also known as BARDA.

BARDA was created in 2006 by President George W. Bush as part of an effort to better prepare the nation against bioterrorism, chemical and nuclear attacks, pandemics, and other public health emergencies that could pose a threat to national security. Casting antibiotic resistance as a threat of that importance was unusual for the United States, but the White House did not want to go through the traditional routes of grants and funding earmarked for basic science. BARDA was designed to be focused. That focus cemented further once it teamed up with the National Institute of Allergy and Infectious Diseases (NIAID). The head of NIAID, Dr. Tony Fauci, was an infectious disease specialist and a longtime member of NIH staff. Fauci had been equally frustrated by Big Pharma's lack of interest in the problem of antibiotic resistance.[11] He wanted to spark some change.

With support from the White House, BARDA and NIAID announced an ambitious proposal on February 16, 2016. The BARDA-NIAID partnership would fund a technology incubator and accelerator, backed by $250 million, to spur innovation and create a pipeline of promising products to tackle antibiotic resistance. The incubator would provide funds to biotech companies but wouldn't own any part of any company that created the drugs. The biotech companies with the most promising products could secure funding from the accelerator, without any requirement to pay it back. But would this bold plan really work?

By this time, Outterson was a professor at Boston University and resident of one of the world's great biotech hubs. Outterson saw the BARDA announcement and immediately called a friend in London to gauge his interest in submitting a proposal. John H. Rex was a senior vice president at AstraZeneca and had worked with Outterson in several meetings during which they drafted policies for the EU. Rex was an MD, a professor, a former NIAID fellow, and, more recently, a pharmaceutical executive. And he was someone who knew the drug-discovery process very well. He had worked on developing new drugs to treat a variety of infections during most of his career. Outterson pitched Rex his idea: creating an innovation hub that would seek the best ideas for antibiotic discovery and development from all over the world. Despite the audacity of the proposal, Rex loved it. He was in.

Which meant that the two men confronted a substantial task. BARDA was not going to fund this plan just because they had an innovative idea. Before it would become a partner in any endeavor, BARDA wanted evidence of a real commitment—and real dollars from academia and industry. To get $250 million, they'd need to do more than write a proposal. So Rex tapped into his network, and the first place he turned to was the Wellcome Trust.

When Henry Wellcome was nine, he saw his town of Garden City, Minnesota, repeatedly attacked by the Sioux. The Sioux were angry over the loss of their ancestral land and further aggravated by the refusal of the US government to pay the compensation they had been promised. The battles were fierce and bloody, and in their aftermath, Wellcome helped his uncle, a medical store owner who assisted in caring for the wounded as well. Wellcome even helped make bullets for the town's defense. But it was the former work that captured his imagination. He had a passion for medicine.

Wellcome went on to study pharmacology, and he became a traveling salesman. In 1880, he traveled across the Atlantic to join a friend, Silas Burroughs, who had formed S M Burroughs and Co., which was importing American-made medicines to the United Kingdom. With the arrival of Burroughs, the two men formed a new company, Burroughs Wellcome & Co., and started delivering a new form of drug—a tablet. At that time, drugs that were available in the UK came in the form of powders or liquids. The tablets were easier and safer to take, because they each contained an exact amount of the drug. The new tablet model was an instant success. The business took off and generated high profits for the company.

After Burroughs died in 1895, Wellcome became the sole head of the company, and under his management Burroughs Wellcome continued to be among the most technologically savvy companies. Henry died in 1936, just a few years before one of his namesake company's worst gaffes: in 1940, two of its chemists visited Oxford to learn about penicillin research, and then politely declined to help the team that was developing the drug.[12]

Henry Wellcome's legacy, however, was preserved for another reason. Before his death, he had created a trust to support biomedical research. And over the years, the Wellcome Trust continued to grow such that, by 1995, it had divested its pharmaceutical assets and was no longer connected to the parent company.

By the time Jeremy Farrar became its director in 2013, the trust was on its way to becoming the largest charity in the United Kingdom for biomedical research (and the second largest in the world). And in Farrar, it had found an ideal director. Farrar had grown up all over the world: he was born in Singapore and raised in New Zealand and Libya.[13] He ended up obtaining his medical degree from University College London and then went to Oxford for his PhD. He was headed for a career in neurology, but just as he was about to finish his academic work, he realized that it was not what he wanted to do. Jeremy pivoted toward infectious diseases. An opportunity in Vietnam brought Farrar back to Southeast Asia, where he spent seventeen years as the head of the Clinical Research Unit of Oxford based in Vietnam. Over those years, Farrar amassed a wealth of experience in public health, infectious diseases, and, most important, the need to combine science with advocacy, and innovation with policy. He took that extensive history with him (as well as having witnessed drug resistance firsthand in Vietnam) when he assumed his position as the head of Wellcome.

In 2016, the Wellcome Trust was approached by John Rex—the accomplished researcher with a history of funding from the trust. He was there to pitch Outterson's idea. The team at Wellcome was intrigued but did not commit to anything during their initial meeting.

Things changed rapidly: Wellcome was in the midst of reviewing how it could better respond to global challenges, and it was working on a new research direction that would focus on bringing in other partners, including innovators and biotech firms. The trust was also eager to back high-risk, high-return projects. Wellcome was leaning toward supporting Outterson's idea, but it wanted a partner to pitch in as well.

By this time, the UK government had already made it clear that antimicrobial resistance was going to be at the top of its domestic, and increasingly global, medical and public health agenda.

As a key step in this agenda, in 2016 the UK government established a new public-private partnership, called the Antimicrobial Resistance Center. Its mandate was to tackle the antibiotic-resistance challenge at a global level. The timing worked out perfectly for Farrar, Rex, and of course Outterson. By the time the grant was due, Outterson had a letter of commitment of $100 million in partner funds from Wellcome and the AMR Center.

On July 28, 2016, BARDA announced that Outterson, Rex, and their team at Boston University were the recipients of the BARDA grant to create an accelerator to tackle antimicrobial resistance. The initiative was called CARB-X[14]—it stands for Combating Antibiotic Resistant Bacteria. There was now real money—hundreds of millions of dollars from the US government and Wellcome to do something different.

The goal of CARB-X is to spur innovation and tackle drug resistance. But it's not an incubator for new companies. It's not even a venture capital firm in a strict sense. It funds preclinical and early development of antibiotics and diagnostics. But there is a catch—it doesn't provide grants for new basic-science ideas. The companies receiving CARB-X money need to have significant funding of their own, and they must have already demonstrated some proof of principle. To incentivize small companies, unlike typical venture capital firms, CARB-X does not own any equity in the company or the final product. It also provides mentorship and guidance to companies in their scientific pursuits and during clinical trials.

CARB-X approached its mandate as if responding to a house on fire. A dozen companies were funded in the first year. They reached the original Year 5 goal in just two years, and the small entrepreneurial team in Boston is now supporting more than forty companies around the world. New funders have joined as well, including the Bill & Melinda Gates Foundation and the government of Germany. Total funding now exceeds a half billion dollars.

Outterson traveled the world in search of good ideas and to convince small biotechs to apply, and he saw real momentum by the end of CARB-X's second year, by which time it had thirty-three projects in seven different countries with five projects in phase I clinical trials. Some focused on new classes of drugs, some on existing targets; there were also vaccines and five projects focusing on new diagnostics. The pace is fast, and expectations are high. But CARB-X is doing what the big companies were unwilling to do.

As Outterson understood all those years earlier when he wrote that footnote in his pioneering report, time is not always on your side. The opposite can be true: you have to run your fastest to try to stay ahead of it.

CHAPTER 30

A THREE-HUNDRED-YEAR-OLD IDEA

Akram Khan* used to work as a truck driver, traveling from one side of the Pakistan-Afghanistan border to the other, hauling goods for masters he didn't know. Sometimes he had to pick things up in the southern port city of Karachi and drive all the way to Peshawar in the northwest, a distance of about nine hundred miles. It would take him two days if he drove fast and slept just a little. After a break in Peshawar, he would cross the border into Afghanistan. All he knew about what he was hauling was that it had something to do with the United States, and it was important. Every now and then, he would hear the word NATO. But he was taught not to ask too many questions. The money was good, he got time off to spend with his family, and nobody was breathing down his neck. But there was a war going on, and whatever was in his truck—and in the other trucks like his—someone wanted to stop. There were drone attacks and, eventually, attacks by helicopters coming from Afghanistan. Locals were getting killed, and authorities acted. The truck traffic stopped, and Khan needed a new job.

Khan needed a job that was reliable. He had to take care of his parents, sister, wife, and four kids. So like many from his village,

_____

*Name changed to protect identity.

he moved down south to Karachi. Like everyone who comes from the part of the country that has had a history of terrorism, he was met with suspicion, but Khan persisted and eventually found a job as a driver for a doctor who knew one of his distant relatives.

In 2018, Khan developed a lingering cough and fever. His cousin told him to get an antibiotic from the nearest pharmacy that would never ask for a prescription. Instead, Khan went to the doctor—the man he worked for. The doctor gave him some painkillers—paracetamol, to be exact—and told him to stay away from antibiotics. The fever subsided with the painkiller, but it came back the next morning. More paracetamol didn't do much. Khan was hoping that the doctor would give him something stronger. The doctor suggested that Khan should get some blood tests done. He sent him to the nearest diagnostic lab.

There are testing labs all over Pakistan, and they make enormous profits. Doctors prescribe a test, and since the hospitals are often underfunded and don't have functioning equipment to run the necessary blood tests, the burden falls on the patient to find a commercial diagnostic lab and pay the requested out-of-pocket fee. Khan went to one of these—knowing full well that the costs of the tests would be deducted from his pay. The results came back three days later. All this time, Khan didn't feel well but still trusted his boss's instincts. He still ran a fever and his throat hurt.

Khan took the test results back to his boss, who looked at them, deemed them inconclusive, and told Akram to get a few *more* tests. The cycle continued for about two weeks, during which Khan lost more money; the lab results remained inconclusive, according to his boss; and Khan didn't get any better. He struggled to keep up with his job, but he was ill, and his boss grew impatient. Khan was fired.

He went to live with his cousin, who gave him an antibiotic. Khan took the antibiotic for a few days and his health improved, and he learned a lesson. He now keeps a stock of antibiotics at home, with a few within easy reach in his pocket. They may be of a lower quality, or adulterated, but Khan doesn't care. He's willing to take his chances—for him it is still better than losing his livelihood.

Khan is one of many people around the world who are failed by a system that is unable to diagnose their illnesses. The tests are expensive, especially for the impoverished, as there is no robust insurance system in place, and, adding insult to injury, the tests are not always accurate. Khan believes that if he had taken an antibiotic when he first got sick, he would have retained his job and his paycheck. He lost his job, and he also lost trust in the system.

Akram Khan's problem, and that of millions of others around the world, could be solved by a technological solution called a rapid diagnostic. A rapid diagnostic is a technology that works at the point of care; it doesn't require heavy machinery, expert staff, or expensive consumables. The world needs rapid diagnostics to accurately, affordably, and rapidly determine the cause of fevers such as the one Khan had—then, once diagnosed, individuals can get the right medicine to cure their ailment. A correct diagnostic would ensure that those who need the antibiotics get them, and those who don't need them are not inclined to keep them in their pockets and contribute to a global problem. Recognizing the necessity of a rapid diagnostic test in antibiotic-resistance screening, a new funding scheme based in the UK is calling on innovators from around the world to take up the challenge. Their model to incentivize is rooted in a three-hundred-year-old idea that changed sea travel and global navigation.

On October 22, 1707, British sailors retreating from Gibraltar to Portsmouth during the War of the Spanish Succession were certain that they were sailing southwest of the English Channel. Instead, they were on course to hit the high rocks off the Isles of Scilly. Nearly two thousand sailors died, making it Britain's worst maritime disaster. The root cause of the accident was the difficulty in determining the true location of the ships and, consequently, where they were heading.[1]

Other similar tragedies continued to mount over the years, so the British Parliament passed a law in 1714 called the Longitude Act:[2] whoever could solve the problem of accurately determining longitude at any point from a ship at sea would be given £20,000, a significant award at the time. Many attempted to solve the puzzle; the winner was John Harrison, an English clockmaker who solved the problem by inventing a marine chronometer to measure longitudes accurately.

Flash forward three hundred years, and another longitude prize was announced. The National Endowment for Science Technology and the Arts (NESTA), a British charity, said that to commemorate the three hundredth anniversary of the work by John Harrison, it was going to look at humanity's greatest challenges once again. It was going to list six of the most pressing issues and ask the public to pick the one that demanded the most attention. The NESTA team chose the following: air travel that wouldn't damage the environment, innovation for sustainable and nutritious food, restoring movement to those with paralysis, creating universal access to clean water, enabling those with dementia to have dignified and independent living, and preventing the rise of antibiotic resistance.

The poll opened in May 2014. On June 24, 2014, Professor Alice Roberts announced the results: "to invent an affordable, accurate, fast and easy-to-use test for bacterial infections that will allow health professionals worldwide to administer the right antibiotics at the right time." The prize this time around was £8 million.

The idea is promising but awfully difficult. Fevers of unknown origin are notoriously hard to diagnose, and even more so at the point of care. While modern methods can detect bacterial infection, the nature of the bacteria, and whether it is susceptible to an antibiotic or not, figuring all of this out *at the point of care*—at the labs frequented by the Akram Khans of the world—currently requires major resources, both human and infrastructural. The goal of the prize is to come up with a way of doing the job accurately, and without the need for either. The prize organizers want to preserve the limited arsenal of antibiotics, and to do so, they want to make sure that only those people who need the drug get the drug—and get it in a timely way.

The world is getting creative in the face of the threat to our health and survival. It has to. If the incentives for the big companies aren't big enough to spark development, then governments and private entities need to step in. The Longitude Prize hasn't been announced yet, but it is another example, another effort to take not just what we have learned from science but what we have learned about its enterprise over the centuries. It is an effort, a desperate call, to harness self-interest, ego, and altruism for the purpose of solving an intractable problem.

CHAPTER 31

SPOONFUL OF SUGAR

A headline caught my eye as I was scrolling the news one day in 2011: "A spoonful of sugar makes the medicine go down."[1] I had heard my children and my wife hum that song endlessly. The headline worked. I clicked to learn more, and I found the news to be about antibiotics and a colleague of mine, James J. Collins.

Collins had been a star athlete during college.[2] He was on the track team at the College of Holy Cross in Worcester, Massachusetts, running cross-country, and by his sophomore year he could run a mile in four minutes and seventeen seconds. He pushed himself to get even closer to four minutes flat.

The excruciating exercise regimen, combined with demanding studies, started to take a toll on Collins's health. During his junior year, he developed a recurring case of strep throat. The doctors would prescribe antibiotics, he would get better, and then, in a few weeks, he would be sick again. He was given erythromycin on thirteen occasions. He felt miserable. His family physician advised him to stop running, telling him that if he didn't, he could cause permanent damage to his heart. Collins took the doctor's advice.

Twenty years later, Collins's mother was taking antibiotics just like her son, and she was also not recuperating in full. Eileen

Collins had been complaining of lower back pain and went to see a chiropractor, who accidentally fractured one of her vertebrae during an adjustment. Collins's mother was then in excruciating pain. At this point, a doctor prescribed her a painkiller, serious enough to be delivered by injection. The needle was probably infected. Soon, she developed a staph infection. Over the next five years, Collins's mother was on and off the antibiotic vancomycin. The doctors would change the drug dose and the regimen, but it made little difference.

Collins's own recurring infection twenty years before, and his mother's more recent infection, likely had a common cause. The infecting bacteria were able to evade the assault from the drug. Not quite resistant, but smart, these bacteria would go into a power-saving mode, or sleep mode, in the presence of the antibiotic. And when the assault was over, these cells would wake up and start multiplying, making the individual sick again. These patient bacteria are called persisters. Strictly speaking, they are not resistant. If they weren't able to drop off to sleep, they would be genetically similar to the cells that are susceptible to the drug. But their ability to turn off means they act as reservoirs of resistance. And in the unrelenting way of evolving bacteria, those that survive multiply, and the next generations tend to have mutations that allow the cells to become resistant. In a few generations, the cells will no longer need to sleep. They will be able to fight the drug through active resistance mechanisms.

Collins took up an interest in persisters by the mid-2000s. His mother's recurring infection came at a time when his team was studying how to wake up the sleeping cells and knock them out. As his own mother's situation revealed, just giving a sick patient more antibiotics wasn't the solution. Something else had to be done. Collins had been following research on persisters and knew that the cells would wake up if their metabolism could be enhanced. If the metabolism could be restored, it would mean that these bacterial cells would get their biochemical machinery back up and running to produce energy to carry out their basic functions, including reproduction. A cell with a functioning metabolism would have all kinds of chemical reactions going on, making it once again vulnerable to antibiotics. This got Jim and his team thinking about sugars. Could adding sugars, either before or concurrently *with* the antibiotic, wake up the persisters and allow them to be targeted?

They tried a number of different sugars, including fructose and glucose, combining them with a number of different antibiotics. They started studying persisters in both Gram-negative (*E. coli*) and Gram-positive (*Staphylococcus aureus*) infections. Most antibiotics did not work when they added sugar. The persisters remained dormant. But then the team tried a class of drugs called aminoglycosides.[3]

Gentamicin, a drug that was discovered in the early 1960s, belongs to this class. Collins and his team did a series of experiments with gentamicin. This time the sugar-drug combination worked. The sugars were taken up by the cells, waking them up, and as a result the drug was effective. Gentamicin is used routinely for urinary tract infections (UTIs), so the team wondered if this sugar-drug combo could work beyond their petri-dish studies. The team decided to test their hypothesis on mice with UTI infections. The drug cleared out the infection and took care of the persisters problem—if they added certain sugars.

The results galvanized Collins and his team, as they thought more about how additives could help antibiotics work better. How antibiotics affect our microbiome is another mystery that his lab is trying to solve. Initially, for scientists, the focus was on killing bacteria—but with recent discoveries underlining the importance of gut bacteria, new products and marketing campaigns, from yogurts to healthy foods, have promoted the idea of preserving, or even strengthening the gut microbiome. This has led to the question, what do antibiotics do to the good bacteria in our gut?

Scientists want to know if the antibiotics truly alter our gut microbiome in ways that are completely irreversible. But Jim was thinking something slightly different—what if we were to use antibiotics and other molecules (such as sugar) that could give the gut bacteria a boost to fight infection? Can we make the gut bacteria do a little bit more to maintain health? Or can we even alter bacteria from within, by taking advantages of natural processes inside bacteria, to make our existing antibiotics work more effectively, even in situations when they don't seem to be effective anymore?

CHAPTER 32

CONFLICT INSIDE THE CELLS

The ruinous war that raged between Iran and Iraq during the 1980s was devastating to both countries, but particularly to Iran. A compulsory draft was sending a near entire generation of young men off to the front lines, from which many never returned. Blackouts and sirens warning of attacks were the norm. The economy came to a halt. Many people started to scout out any opportunities to escape.

Houra Merrikh's family was among them.[1] Living in central Iran, not far from Gorji Street, named after Merrikh's mother, they were affluent, respected, and powerful. But the war changed everything. In 1983, Merrikh's brother was about to turn fourteen, and no one wanted him to be drafted into the revolutionary army. The family had applied for a US green card, thanks to an uncle who was living in the United States. The immigration office handling their case at the US State Department, while seemingly helpful, was in no hurry. And any further delay in their leaving Iran could mean that Houra's brother might get drafted and be used as a human shield in the great cause of the current holy war.

The family moved to Turkey. They landed in Istanbul in August 1985. There, they reinitiated the process of getting a green card. Every six months they would get a hearing with the US consulate in Istanbul, and every six months they were told that the arrival of the card was imminent. This process continued for thirteen years. After trying unsuccessfully to settle in Iran, Houra's family relocated to Ankara in 1988. The family was now struggling. All this time, they had been waiting for their green card. In 1989, Houra's brother graduated high school and was getting ready to go to North Cyprus to study engineering. Merrikh's father returned to Iran first, and then Merrikh and her mother followed him there. There they found things far more difficult. After just nine months, they both moved to North Cyprus to live with her brother in a small apartment shared with several university students.

They had little money. The house they were living in was riddled with bullet holes, a reminder of the Turkish invasion of Cyprus in 1974. With few choices, Merrikh and her mother struck an arrangement whereby they would get free housing in return for cooking and doing laundry for the university students. Merrikh was only nine years old at the time. She and her mother took turns in the kitchen, but her mother was often ill, struggling with mental illness. Eventually, her mother was diagnosed with bipolar disorder, and Merrikh's parents divorced. Merrikh and her brother had to take care of themselves as well as their mother.

It was no longer a cause for excitement when their green card came. Years had passed, and it was now cause for concern. The green card was not going to bring about a future of family reunion—instead it would further break the fragile family apart. Merrikh's brother was over twenty-one, so he no longer qualified. Her parents were now divorced, so her father did not qualify either. Her mother's struggles with mental illness, together with a lack of financial or family support, prevented her from moving as well. Merrikh was the only one who could leave.

Houra Merrikh was just out of high school and working as a hotel receptionist, waiting for her medical tests to be completed so she could finally go to the United States. She was planning on living with her aunt in Seattle. But when she called her aunt to finalize the details, her aunt refused to help her. The aunt's own family was struggling too much to take on the additional responsibility. Merrikh was very close to the end of her rope.

Then there was an unexpected turn of events. An Iranian family living in Texas, whom Merrikh had never met before, were staying at the hotel where Merrikh worked. They took an interest in one another, and eventually the family asked Merrikh about her plans with the green card, expecting to hear all about how excited she was. Instead, they heard despair in Merrikh's voice. The Alavis family told Houra that she could stay with them. Merrikh left for Texas a week later.

Merrikh landed in Dallas–Fort Worth and immediately turned to the necessities. She secured a driver's license, a Social Security card, and a new job at an ice-cream shop that paid her $6.25 an hour. She rented her own apartment, modest to be sure, but also set in motion the steps that brought her ill mother to the United States a couple of months later. She hustled and got enrolled in a local community college. She moved from Arlington to Austin, and then to Houston, always maintaining a perfect GPA. And at the University of Houston, Merrikh fell in love with biochemistry. She completed her bachelor's degree magna cum laude in biophysics and biochemistry, while receiving several prestigious scholarships and grants.

Merrikh's love of science took her from Houston to Boston, landing first at Boston University as a research technician and then at Brandeis for graduate school—where she worked on bacteria and DNA. She excelled, working with Dr. Michael Rosbash, who won the Nobel Prize in 2017, and later with Dr. Susan Lovett, with whom she did her thesis work, completing it in a year and a half. Soon after receiving her PhD, she moved to MIT for postdoctoral training and quickly landed a coveted tenure-track job across the country at the University of Washington, Seattle, where she had once dreamed of starting off her life in the United States.

By the time she finally reached Seattle, Merrikh was no stranger to conflict, complexity, and collisions. Her lab focused on what happens when existing systems and structures collide. Of course, she and her colleagues did so by looking inside the cell—at two fundamental biological processes: transcription and DNA replication.

Transcription is the first step in gene expression. It is the process by which the double-helix DNA copies its information into RNA. RNA, or ribonucleic acid, is present in all living things, and its primary purpose is to convert information contained in the DNA to make proteins. Proteins are the workhorses of a cell that do most of the work and control how tissues and organs are formed. Transcription is the first step in a process that leads to the formation of a protein. Replication, as the name suggests, creates another copy of DNA that ensures the new daughter cells are like their parent.

The two processes sometimes move along the same twisty double-stranded ladder of DNA. Think of them as two trains moving on the same track—sometimes they'll move in the same direction and sometimes they might collide with each other. These collisions can lead to mutations—both favorable and unfavorable.

Mutations, of course, are central to evolution. But Merrikh also appreciated that cells have other mechanisms that can accelerate evolution. Merrikh wondered if we could use this information about collisions to reverse the clock. Could evolution be slowed down? This question led her to another: What if, by slowing down evolution, we could stop bacteria from developing resistance? During cellular processes, like replication or transcription, DNA might get damaged and need to be repaired. This repair is carried out by proteins that fix this damage. Mfd is one such protein to fix DNA damage.[2] Merrikh's lab had shown that Mfd can also increase the rate at which mutations occur.[3] What if this protein was removed? Could doing so keep bacterial cells from becoming too sophisticated, and, if so, would they respond to antibiotics?

By this point, Merrikh was a rising star in her field and was awarded the Vilcek Prize—a coveted award given to immigrants who demonstrate exceptional creativity. After she received the prize, Merrikh was invited to meet with the dean of her school, where she was asked about the next steps in her work. Merrikh pitched her new ideas. Blocking evolution was both creative and seemingly crazy, which is precisely the sort of idea that the NIH was not interested in funding. The dean noted both the unusual idea and Merrikh's frustration over a lack of funding.

Three days later, she received an email from the Bill & Melinda Gates Foundation with a simple message: Bill wants to see you on Wednesday.

In the small meeting arranged for that Wednesday, there were public health workers from Africa and India who were talking about the impact AMR was having on their own communities. The stories from the front line were devastating. Though they were originating far from Iran and Turkey, they were nonetheless all too familiar to Merrikh, reminding her of her personal struggles with poverty, despair, and hopelessness.

The meeting with Gates and the AMR group was life changing for Merrikh, and she was now convinced that her crazy idea really did matter—and really could work. And she had funding. The Gates Foundation had decided to back her.

Very carefully, Merrikh and her team targeted Mfd and studied the effects of antibiotics on disease-causing bacteria that had Mfd, versus the bacteria that did not have Mfd.[4] They worked with *Salmonella* and with TB. The results were striking—the bacterial cells that did not have Mfd were nearly a thousand times less likely to become drug resistant. The results created a buzz in the scientific community. Could evolution truly be stopped by a drug? Can we give two drugs to a patient with an antibiotic-resistant infection—one antibiotic and one "evolution stopper"?

Merrikh is on the hunt to figure out how to make an evolution-stopping drug that could be used to tackle AMR. She knows that the road ahead is not going to be easy or straight. FDA approval for an evolution stopper would be unprecedented. Human trials will mean ethical issues that will need to be studied in detail—but the potential upside to this unparalleled approach is in the impact it could have on poor communities around the world, where diseases such as TB remain stubbornly difficult to treat.

Merrikh's path to a solution is an inspiring example of human ingenuity and success in the face of formidable challenges. The Iranian family in Texas, the Gates Foundation, and all of the many people and organizations who have come together to support Houra Merrikh—all enablers of human ingenuity—are also part of this tale of scientific triumph and hope. Despite long odds and extraordinarily difficult circumstances, some genius persists, and it may finally tip the scales in our contest with bacteria.

CHAPTER 33

SECURITY OR SERVICE?

On October 15, 2017, a fifty-year-old doctor took the stage in front of more than a thousand people in Berlin. The meeting is one of the biggest annual events in global health—the World Health Summit—and has been held annually in Germany's capital city since 2008. Sitting in the large conference room were the ministers of health from Germany and Portugal, as well as diplomats and senior health officials from the WHO, the United States, the European Union, Japan, and elsewhere. Heads of major pharmaceutical companies of Europe were also facing the stage. Almost every organization that plays a part in solving the problems associated with global public health had a representative in that hall.

Dr. Joanne Liu, the head of Médecins Sans Frontières International (MSF), or Doctors Without Borders, at the time, was there to deliver a message calling out the hypocrisy of Western governments in fighting disease.

By training, Liu is a pediatrician. The daughter of Chinese immigrants who ran a Chinese restaurant in a small town in Quebec, she had gone to a French elementary school and speaks English with a French-Canadian accent. Some in the audience knew her; others were learning about her work for the first time. Liu spoke calmly, thinking carefully about each word. She had ten minutes and intended to make her point. She focused on Ebola, and her message was simple.

Ebola became a priority only when the United States and Europe determined it was a national security issue. Prior to that, the deaths of nearly eleven thousand people in West Africa were just another statistic of a tragedy occurring in a far-off land with a broken health system and poor people. Giving another example, Liu talked about Yemen, which was left to collapse due to the recent war. One after another, her examples—Africa, the Middle East, impoverished areas in South Asia—reinforced her point.

Her final shot was aimed squarely at the ministers and bureaucrats sitting in front of her, and it made them uncomfortable. She argued against the wealthy world's habit of approaching health emergencies only through the lens of their own national security. How, she asked, could national security interests outweigh the health and well-being of people struggling to survive? There can be no global health when nations with the means to act put their security over the lives of people in other countries. Do not, she implored, allow the summit to be like the previous events: "sterile conversations that amount to nothing."

Joanne Liu got a standing ovation.

Liu vividly recalls why she decided to become a doctor.[1] It wasn't her first passion. That was hockey. But her life changed when she read *The Plague* by Albert Camus. A specific line stuck with her, and it is when the protagonist says, "I am still not used to seeing people die."

Working for MSF was her dream job. In 1996, she got that opportunity, and Liu started her career with MSF in Mauritania. She quit three months later. At that time, she thought that MSF policies were more about doing something—but not necessarily about doing the right thing. Her dream was reduced to bitter disappointment. Despite her frustrations, she still believed in the mission of MSF, and so she came back to the group and worked in Sri Lanka, Kenya, Palestine, Haiti, and Afghanistan, even as she continued to practice as an academic pediatrician at McGill University in Montreal. In 2013, she ran for the top position at MSF International. She was elected. Antibiotic resistance had been on her agenda for years, and ever since she became the president of MSF, the organization has shared that passion.[2]

MSF treats more patients suffering conflict-related trauma than any other organization. This means their staff members regularly encounter infections and antibiotic use. Historically, MSF has stayed on the sidelines in the resistance debates, continuing their "best practice" approach—which included administering broad spectrum antibiotics to prevent infections. These antibiotics aren't specific to a particular ailment but go after a variety of infection-causing bacteria. The use of broad spectrum antibiotics is concerning because these drugs adversely affect our gut flora. Also, any resistance that develops against broad spectrum antibiotics would mean that a whole arsenal of antibiotics would become useless, compared to resistance developed against a specific antibiotic. But MSF argued that their mission was to save lives—and they had to do so on a shoestring budget and in the shortest possible amount of time.

Testing and creating patient-specific treatment plans was simply not viable. Running long, expensive, and extensive tests was not an option either. Then there is the risk of developing resistance due to poor postoperative care. MSF can do infection control in their hospitals, but they are unable to take care of all postoperative care at home, especially in the conflict-ridden, impoverished places where they do their most important work. They want to change the global culture of antibiotic use, but given the interests of global powers, it's not an easy task. International and domestic security is increasingly becoming the lens through which antibiotic resistance is viewed, and it's making many in MSF uncomfortable.

An example is the Global Health Security Agenda (GHSA), launched in 2014. The United States has taken on a leadership role in creating and expanding its framework, with the idea being to promote a national and international security approach to the problem. Its proponents argue that antibiotic resistance is a threat to economies, people, and hence national security. It should be treated as the world would treat the emerging threat of a pandemic.

The argument has won support in the corridors of power and funding from the United States, and other European governments have come along with it, on the order of billions of dollars. But not everyone is excited. The language often used in advancing the global health security agenda has warlike connotations, and while the goal is to tackle resistant infections, the term *security* makes public health professionals nervous. Those who work in war zones worry that this language is not conducive to creating a global campaign that views all lives equally. Will the agenda extend to protect the lives of civilians who may live in "enemy territory"? For many, the concern can be simply expressed. Securitizing health care seems too easy a slippery slope toward weaponizing it. In the safe cities of the developed world, security sounds comforting; in the developing world, often overseen by dictators and autocrats, it can take on a far more threatening connotation.

Joanne Liu wanted to change the role MSF plays in the international debates on antibiotic resistance. She no longer wished for MSF to be on the sidelines of the debate. She is among those who are skeptical and concerned about increasing the securitization of global health. She's worried that under the guise of security, only the rich countries will get to benefit and the poorest will, as is so often the case, be left to suffer. A better way, she argues, reflecting on decades of experience in war zones and humanitarian crises, is to ensure the dignity of every person, in every country, regardless of that country's strategic importance to the United States and Europe. Liu wants the world to confront global conflict, which she sees as one of the biggest blind spots in antibiotic resistance. Unafraid to rattle the nerves of those in power, Liu is directly addressing one of the systemic reasons for resistance's spread and one that is too often swept under the rug.

Through our incessant wars and conflicts—be they sectarian, tribal, national, or regional—we're doing bacteria's work. We are providing bacteria with the medium in which they can thrive, adapt, resist, and affect all of us—even if we are thousands of miles away from the conflict zone. The only way to ensure that we are not in a perpetual state of catch-up is to ensure that each one of us is healthy—and that doesn't happen through the barrel of a gun.

CHAPTER 34

ONE WORLD, ONE HEALTH

Steve Osofsky was seven years old when he looked the white rhino in the eyes.[1] Osofsky was at the Catskill Game Farm, only a couple of hours from his home, but to a young boy, it felt like it was a world away. No doubt it felt similar to the rhino.

Osofsky's dad told him that the rhino was from South Africa, a country on the other end of the world from New York. And yet there they were, however improbably, looking into each other's eyes, underscored by the simple thought that perhaps their lives were more entwined than humanity normally admits to. That encounter was transformative; young Osofsky decided that he wanted to become a wildlife veterinarian.

During high school, college, and vet school, Osofsky gained a variety of important experiences working with domestic and wild animals. He worked with Florida panthers in the Everglades and with elephants in Kenya. Wherever he went in the world, the common thread in his mind was conservation.

After he received his veterinary medicine degree, he completed a one-year small animal medicine and surgery internship. Then an opportunity arose in the small rural town of Glen Rose, Texas—about seventy-five miles southwest of Dallas—at a place called Fossil Rim Wildlife Center. The center, covering eighteen hundred acres and open to the public since 1984, is dedicated to protecting endangered species. Osofsky threw himself into the work, and his experience at Fossil Rim opened up new possibilities. He started sending out his résumé to countries all over Africa, asking if they needed a wildlife veterinarian. The government of Botswana wrote back to say that they might have an opening.

Osofsky jumped at the opportunity, packed his bags, and became the Botswana Department of Wildlife and National Parks' first wildlife vet in 1992. His home base was in the capital of Gaborone, but he was constantly traveling around the large, sparsely populated country, and into its national parks and game reserves.

Foot-and-mouth disease was a particular challenge facing Botswana at the time. It could be devastating for livestock farmers, for any evidence of the disease in their cattle prevented their access to the world's beef markets. The natural vector for foot-and-mouth disease was the African buffalo, which can pass the virus to cattle. The historical methods that countries such as Botswana have adopted to protect their livestock from the disease have been administering vaccinations and building vast fences to keep buffalo out. These veterinary cordon fences, however, have harmed the migratory wildlife since they were first constructed in the late 1950s. Decades later, the problems had not been solved. Like many before him, Osofsky saw no obvious solution.

He returned to Texas in 1994—this time taking on a leadership position at the Fossil Rim Center. A couple of years later, he earned an American Academy for the Advancement of Science (AAAS) Fellowship at the US Agency for International Development in Washington, DC, another life-changing opportunity. He subsequently moved into the wildlife conservation nonprofit sector, and in 2003 launched the Animal & Human Health for the Environment and Development (AHEAD) program at the Wildlife Conservation Society (WCS).

He joined the WCS at the invitation of a man named William (Billy) Karesh. Karesh had had a remarkable journey himself.[2] Growing up in Charleston, South Carolina, he found animals fascinating. His home was a shelter for orphaned blue jays, squirrels, and raccoons. He created his own "soft release" program for the animals that he brought home every summer. In college he struggled to find the right major and went from business to engineering, until the mother of a friend of his encouraged him to work with animals, his real passion. Karesh got his undergraduate degree in biology from Clemson University and followed it up with a veterinary degree from the University of Georgia. After veterinary school, a series of internships and jobs around the world brought him to the Bronx Zoo, where he now heads the WCS.

In the early 2000s, Karesh coined the phrase *One Health* as a new way to think about animal and human health from a single lens. Osofsky, who was now working for Billy, along their colleague Bob Cook, decided to put together the first One World, One Health conference in 2004 at the Rockefeller University.[3] The timing for the conference, linking human, animal, and environmental health and stewardship, couldn't have been better. Diseases such as avian influenza, Ebola, and chronic wasting disease drove home the point that it was becoming more and more difficult to view human and animal health from two disparate viewpoints.

Steve was excited, but he wanted to end the conference with a big idea, something that would create a foundation for action and amplify the application of One World, One Health. He ended up drafting the Manhattan Principles on One World, One Health, which were discussed and then adopted at the forum.

The principles are a series of statements that urge the world's leaders, civil society, and the global health community to recognize the interconnectedness of our world.

There are twelve principles, including a call for recognizing "the essential link between human, domestic animal and wildlife health and the threat disease poses to people, their food supplies and economies, and the biodiversity essential to maintaining the healthy environments and functioning ecosystems we all require." Other principles call for increased investments in animal and human health infrastructure, a demand for international collaboration and increased awareness through education. Some are more specific, including an effort to reduce the demand for bushmeat and a call for restricting mass culling of free-ranging wildlife unless there is "a multidisciplinary, international scientific consensus that a wildlife population poses an urgent, significant threat to human health, food security, or wildlife health more broadly." But missing from the list was any mention of antimicrobial resistance.

A series of global epidemics meant that One Health and the Manhattan Principles caught on quickly. There were periodic reports of avian flu viruses coming from East Asia. The public was afraid of new diseases jumping to humans from birds, pigs, and cattle. The Food and Agriculture Organization (FAO) saw an opportunity to make animal health and welfare part of the global debate on antibiotic resistance and became an early advocate of One Health, and so did the World Organization for Animal Health (OIE). The CDC also created a One Health office in 2009. For the next several years, pandemic preparedness became an area of focus, including surveillance, diagnostics, and containment. Antibiotic resistance would occasionally surface in conversations, but it was far from the center of the debate. In 2015, this would completely change.

A paper published in November 2015 made it clear that the widespread use of colistin on pig farms in China was creating resistance in bacteria that was affecting both animal and human health.[4] Other reports of similar cases followed in quick succession: on poultry farms in India, in Southeast Asia, and as far as industrial livestock in Europe. What Stuart Levy had seen years ago in the United States and Witte in East Germany—the jumping of resistance from animals to humans—is now happening at a global scale. Small DNA units, plasmids, can go from one bacterial species to another, transforming ordinary bugs into superbugs. These resistant bacteria are free to travel, going from the farm to humans through meat and contaminated waterways. Human and animal health are interconnected in a myriad ways—and solutions have to be cognizant of both. No longer can we look at the health of animals and humans in silos. The One Health approach offers one of the best ways forward toward finding a viable solution.

CHAPTER 35

BANKERS, DOCTORS, AND DIPLOMATS

The most oft-told account of scientific discovery and technological innovation tends to be disingenuous. The story, told through books and legends, movies and plays, has a lone-wolf scientist as the protagonist. This character—almost always a male—through serendipity, creativity, and sheer force of will, makes a ground-breaking contribution. Almost always lost in this rendering is another character, one who stays in the background: the enabler of the discovery. It could be a financial backer, a visionary politician, the willing public, or the scientific enterprise writ large. Scientific genius and a good deal of luck are almost always necessary, but also almost never sufficient. In the grand scheme of discovery, a large cast of characters and historically rich context are needed. Tackling antibiotic resistance has always been no different.

Since its establishment in 1885, no woman had ever been appointed the chief medical officer (CMO) of the United Kingdom. Sally Davies would be the first, assuming the post of the country's top public health officer in 2010.[1] The CMO's task, most broadly construed, is to advise the government on the preservation of the community's health. Her predecessor, Sir Liam Donaldson, for example, worked to decrease Britain's high rates of alcoholism and public smoking.

Davies grew up in a family of academics—her mother a scientist and her father a theologian—among whom ethical and moral questions were debated regularly. She went on to study medicine, receiving degrees from the University of Manchester and the University of London. In 2012, soon after becoming the CMO, she stunned her colleagues when she announced her decision to take on antimicrobial resistance. The old guard of the British public health civil service was unhappy. Of all the problems that Davies could focus the nation on, from diabetes to mental health to cardiovascular disease, antimicrobial resistance seemed like an odd choice at best and a distraction from more pressing concerns at worst. There were murmurs as to whether the CMO was even in touch with the realities of British public health.

To Sally Davies, who by then had been knighted and was now Dame Sally Davies, her choice of focus was not just about a present danger, it was also about connecting the present with the future and the past. Microbiology, after all, had once been the crown jewel of British biology. To get the nation's attention, Davies issued a report on antimicrobial resistance. Davies also knew from the outset that a mere report on antibiotic resistance was not going to be enough to get the national and global attention that the problem deserved. That would require the efforts of someone far more prominent than the CMO. That would require the British prime minister, David Cameron.

Davies requested a meeting with the prime minister and was told he would see her on March 19, 2014. The meeting did not last long, and by the end of it she knew that Cameron was convinced. Something needed to be done.

Four months later, in an interview, David Cameron announced the following: "This great British discovery has kept our families safe for decades, while saving billions of lives around the world. But that protection is at risk as never before."[2] The discovery Cameron was talking about was antibiotics, recalling the path-breaking work of the so-called Dunn School of Oxford. He had just commissioned a report on the threat and future of antibiotics, which on its publication would be cited thousands of times and become the gold standard for the world's response to resistance over the next five years. Heading that commission and overseeing the report was not a scientist or a doctor, but an economist and financier: Lord Jim O'Neill.

You could argue that Jim O'Neill is lucky.[3] Consider that on September 9, 2001, he was conducting meetings in New York City in one of the Twin Towers. He was supposed to stay on for a couple more days, discussing important economic issues during a conference for business economists, but he decided to leave early, flying home on the eve of September 11. You could also argue that he was prescient. Trained as an economist, his area of expertise was China, the Asian economic crisis of the late 1990s, and the mounting challenges of globalization, all of which led him to write a paper in November 2001 entitled "Building Better Global Economic BRICs."[4] The term BRIC was an acronym for Brazil, Russia, India, and China, and O'Neill's use of it coined the term, which is now part of the standard economic jargon that refers not just to those four countries but to all emerging economies. But perhaps more than anything, you could argue that Jim O'Neill was someone who got people's attention.

In May 2014, Jim got a call from a senior member of the British Treasury, who proposed that he take on a new role, one that was going to be launched by Prime Minister David Cameron. The task? To lead a comprehensive global review on antimicrobial resistance. Jim had never even heard of the issue. He went home to mull over the proposition and, persuaded by his family to take on this challenge, Jim agreed.

A month later, Jim O'Neill told Sally Davies that he was going to do things differently. He was going to take a crack at the problem from the lens of global economics, an area he knew well.

Over two years of countless meetings, data gathering, and number crunching marathons, Jim and his small team came up with what they called their ten commandments to turn the tide. These included a massive global public awareness campaign, improvements in hygiene and the prevention of infection, reduction in the use of antimicrobials in agriculture and the livestock industry, global surveillance of drug resistance and consumption, development of rapid diagnostics, development of vaccines and alternatives, launch of an innovation fund for developing new diagnostics and drugs, improvement of the incentives for research and development in making new drugs (and improving existing ones), and creation of a global coalition for action. He also advocated for improving the working conditions, pay, and public recognition of people working in infectious diseases.

The findings of O'Neill's commission have now become part of nearly all discussions on the challenge of antimicrobial resistance.[5] What has stuck with the world, however, is not the list of the ten commandments, but an estimate from Jim and his team. If nothing changes, and we continue on the path we're now on, by 2050 the world will lose 10 million people a year, every year, to resistant infections. That is equal to losing nearly all of the inhabitants of New York City and Chicago every year.[6] The O'Neill report was not without its critics,[7] but by the fall of 2016 it had started an international conversation.

Davies used O'Neill's report to push for global change. Her crowning moment came on September 21, 2016, at the UN Headquarters. It was only the fourth time in the history of the UN that the international body had a high-level meeting on health. It was to consider a resolution to combat the growing threat of antimicrobial resistance. All 193 member countries voted in favor of the resolution. It had been a long road, with countless meetings and efforts by politicians and staff members, doctors and activists, in countries around the world. Davies's decision to choose antimicrobial resistance as the problem of our era had been vindicated.

More important, in this telling of the story of scientific advance—in which England's first female chief medical officer secured the support and public interest of a prime minister, who in turn tasked an economist with solving a global health problem—we can spy reasons for hope. When humankind, rather than any individual or even any individual nation, confronts a threat that plausibly risks the lives of 10 million people annually, trusting the brilliance of a lone scientist as the only hope is not only unwise, it is ahistorical. Scientists and innovators with the brightest and boldest ideas need our support through sustained funding, but so do social scientists, economists, humanists, policy makers, and public health practitioners. Our collective future needs to pay attention to all the relevant characters in what is a human drama if it is to have a happy ending. For the protagonist is, in fact, us.

EPILOGUE

Biographies are rarely about the future. But antimicrobial resistance does have a future, one that is going to affect the way we get to live and die. The potential doomsday scenario of tens of millions dead annually is real, but so are the hopeful developments of the last few years. On the technical side, there is promise in vaccines[1] and phage therapies.[2] On the economic front, ideas are being proposed that incentivize pharmaceutical companies to commit themselves to research and development.[3] There is a new sense of urgency within the WHO to improve surveillance and empower all countries, rich and poor, large and small. Institutions such as the Pew Charitable Trusts are using a combination of data gathering, information sharing, and awareness to highlight the risks and possibilities associated with the problem of resistance.[4] The Bureau of Investigative Journalism has created its own project to highlight the problem through investigative reporting from the epicenters of resistance and infection.[5]

Is all of this enough to turn the tide?

We should be collectively terrified that there is no obvious answer to that question. The drugs in the pipeline may or may not succeed. Markets and investors are ruthless, and they are likely to park their money where there are maximum financial returns, not necessarily where there is the maximum benefit to humanity. Phages are promising, but there are few large-scale clinical trials to support their promise. Vaccines are not going to be able to replace all antibiotics. And the anti-vaccination movement, which comes in different forms around the globe, continues to create challenges as well. International awareness campaigns are welcome and well intended, but they rarely reach the small farmers in rural communities in places like Burkina Faso or Bolivia. National action plans, many of which were created after the UN resolution in 2017, are just plans on paper. They will need money and political will, and right now too many officials are sitting in the offices of various ministries of health with no clear plans for implementation.

Also concerning is the populist reaction to globalization, which is seen solely through an economic lens rather than through one of antimicrobial resistance. A desire to decrease international engagement, support nativism, erect high walls, and punish those who disagree with "my-nation-first" reasoning has made it harder to create international partnerships. Gili Regev, an infectious disease expert in Israel who has been working with Palestinian doctors and patients for over a decade to address the issue of drug resistance in poor communities in Palestine, is unable to continue her work due to the US desire to cut funding for any programs that support the Palestinians. Hospitals that have come under attack during the Saudi-led conflict in Yemen, or by Taliban in Afghanistan, or by armed gangs in the Congo, erode decades of effort to contain infection and manage antimicrobial resistance. The bullets and bombs used in such conflicts are triply fatal. They destroy the lives of their immediate victims, lay waste to critical health infrastructure, and aid in the spread of superbugs that threaten humanity itself.

Far from acting to help itself, humanity seems currently to be aiding the superbugs.

Bacteria will continue to do what they have done since the dawn of life—evolve, adapt, and get ready for the next battle for survival. Our actions are helping them acquire a better arsenal at a faster rate than they probably would have on their own. But despite the challenges and frustrations, in the hundreds of interviews that I conducted for this book, there was a sense of optimism about the future. That optimism stems from a belief in human ingenuity, the vast reserves of natural treasures that are untapped, and the power of coming together. That optimism is also predicated on two things: a commitment to peace, and a desire to care for all people—everywhere.

ACKNOWLEDGMENTS

The opportunity to work on this book made a profound impact on my life. On one hand, it opened new avenues for research and inquiry; on another, it gave me the privilege to meet some of the most dedicated people I have ever come across. This book would not have been possible without the incredible support and generosity of colleagues and friends. In Boston, Scott Podolsky was always available to share his knowledge, wisdom, and resources, and had answers to all of my questions—even when they did not make any sense. In Oxford, I got to meet Claas Kirchhelle, one of the most brilliant historians of science and medicine I have ever met. Claas, like Scott, helped me throughout the process. My discussions with Robert Bud in London were instrumental in helping me understand the early days of penicillin and the subsequent antibiotic policy in both the human and veterinary sector. In Oslo, Anne Helene Kvieim Lie was there to help facilitate my research in Norway in particular, and in Scandinavia in general. I owe a great debt to her for the material that shaped multiple chapters in this book. In Russia, Anna Eremeeva helped me a great deal with material on Zinaida Ermolieva. In Saint Petersburg, Sergey Sidorenko arranged for meetings with Yakov Gall and others. In Moscow, Olga Efremekova facilitated my work at the Gause Institute in particular, and Moscow in general. During the early days, Andre Sobocinski, in the US Navy, helped me navigate military archives and connected me with colleagues in the vast universe of military medicine. Daniel

Berman at NESTA in the UK helped me understand the vision behind the Longitude Prize. I am also grateful for the time and candid insights of colleagues at the WHO, the FAO, and other international agencies that are working tirelessly to address the issue of drug-resistant infections, despite all the political and financial odds. I will always remember the help I received from the librarians and archivists in Boston, Berlin, Geneva, London, Moscow, Tokyo, and Washington, DC.

I wrote part of this book in the beautiful environment of Boston Athenaeum. Colleagues and staff there have been generous and gracious with their time and support.

My agent, Michelle Tessler, has been there from the start, guiding me at every step of the way. At Harper Wave, Karen Rinaldi and Rebecca Raskin were a delight to work with. Karen in particular was always encouraging, supportive, and most resourceful. I am also very grateful to Amanda Moon and Thomas LeBien—both of whom helped me sharpen the arguments and create a narrative that made this book possible. Amanda and Thomas are kind, generous, and above all exceptionally gifted. It was a sheer pleasure to work with them.

My colleagues at the Howard Hughes Medical Institute, in particular Sean Carroll, David Asai, and Sarah Simmons, were immensely supportive of this project and helped me in every possible way. Sean, a celebrated science writer and scholar par excellence, in particular helped me a lot when this book was just an idea.

I also owe special thanks to Dr. Carly Ching and Dr. Sam Orubu in my research group, both of whom read the manuscript several times and provided candid feedback that made the manuscript richer in so many ways.

My sisters, Rabia and Fakiha, and their husbands, Umar and Hamza, respectively, and my sister-in-law, Shaista, have been a constant support of encouragement and support. My nieces and nephews are a boundless source of joy. My brother, Qasim, continues to inspire me in myriad ways through his depth of knowledge and intellect. While our academic fields are different, his rigorous approach and thorough scholarship continue to be a shining light for my own work. I am also grateful for the constant support of my in-laws, in particular my mother-in-law, Tarannum Siddiqi.

My son, Rahem, and my daughter, Samah, brighten our home every day. Their wit, charming smiles, and infectious laughter made our home the best place to write this book.

My biggest gratitude, however, is to my wife, Afreen. There are no words—in English or in any other language that I know—to thank her. She is not just my partner but also my best friend and someone who has helped me at every single step of this work. From the moment the idea for this book was conceived, to the email with the final edits, she has been there for me. Her never-ending support, kindness, insight, and input enabled me to write this book. Without her, this book would not have been possible.

NOTES

PROLOGUE

1   Lei Chen, Randall Todd, Julia Kiehlbauch, Maroya Walters, and Alexander Kallen, "Notes from the Field: Pan-Resistant New Delhi Metallo-Beta-Lactamase-Producing *Klebsiella pneumoniae*—Washoe County, Nevada, 2016," *Morbidity and Mortality Weekly Report* 66, no. 1 (2017): 33.

2   Helen Branswell, "Can a Flu Shot Wear Off If You Get It Too Early? Perhaps, Scientists Say," Stat News, January 12, 2017.

3   Neil Gupta, Brandi M. Limbago, Jean B. Patel, and Alexander J. Kallen, "Carbapenem-Resistant *Enterobacteriaceae*: Epidemiology and Prevention," *Clinical Infectious Diseases* 53, no. 1 (2011): 60–67.

4   L. S. Tzouvelekis, A. Markogiannakis, M. Psichogiou, P. T. Tassios, and G. L. Daikos, "Carbapenemases in *Klebsiella pneumoniae* and Other *Enterobacteriaceae*: An Evolving Crisis of Global Dimensions," *Clinical Microbiology Reviews* 25, no. 4 (2012): 682–707.

5   Jesse T. Jacob, Eili Klein, Ramanan Laxminarayan, Zintars Beldavs, Ruth Lynfield, Alexander J. Kallen, Philip Ricks et al., "Vital Signs: Carbapenem-Resistant *Enterobacteriaceae*," *Morbidity and Mortality Weekly Report* 62, no. 9 (2013): 165.

6   Kevin Chatham-Stephens, Felicita Medalla, Michael Hughes, Grace D. Appiah, Rachael D. Aubert, Hayat Caidi, Kristina M. Angelo et al., "Emergence of Extensively Drug-Resistant Salmonella Typhi Infections Among Travelers to or from Pakistan—United States, 2016–2018," *Morbidity and Mortality Weekly Report* 68, no. 1 (2019): 11.

7   Emily Baumgaertner, "Doctors Battle Drug-Resistant Typhoid Outbreak," *New York Times*, April 13, 2018.

8   CDC report available at https://wwwnc.cdc.gov/travel/notices/watch/xdr-typhoid-fever-pakistan.

9   Jason P. Burnham, Margaret A. Olsen, and Marin H. Kollef, "Re-estimating Annual Deaths Due to Multidrug-Resistant Organism Infections," *Infection Control & Hospital Epidemiology* 40, no. 1 (2019): 112–13; Centers for Disease Control and Prevention, "More People in the United States Dying from

Antibiotic-Resistant Infections than Previously Estimated," CDC Newsroom, November 13, 2019, https://www.cdc.gov/media/releases/2019/p1113-antibiotic-resistant.html.

10 Susan Brink, NPR, January 17, 2017. https://www.npr.org/sections/goatsandsoda/2017/01/17/510227493/a-superbug-that-resisted-26-antibiotics.

CHAPTER 1: WHAT WE'RE UP AGAINST

1 Alison Abbott, "Scientists Bust Myth That Our Bodies Have More Bacteria Than Human Cells," *Nature* 10 (2016).

2 Dorothy H. Crawford, *Deadly Companions: How Microbes Shaped Our History* (Oxford: Oxford University Press, 2007).

3 Ibid.

4 Christoph A. Thaiss, Niv Zmora, Maayan Levy, and Eran Elinav, "The Microbiome and Innate Immunity," *Nature* 535, no. 7610 (2016): 65.

5 William Rosen, *Miracle Cure: The Creation of Antibiotics and the Birth of Modern Medicine* (New York: Penguin, 2017).

6 Ibid.

7 For a general background on antibiotic resistance mechanisms, see ReACT guide, available at https://www.reactgroup.org/toolbox/understand/antibiotic-resistance/resistance-mechanisms-in-bacteria/. A more detailed account is available at Jessica M. A. Blair, Mark A. Webber, Alison J. Baylay, David O. Ogbolu, and Laura J. V. Piddock, "Molecular Mechanisms of Antibiotic Resistance," *Nature Reviews Microbiology* 13, no. 1 (2015): 4.

8 For a discussion of MRSA origin and its potential threat, see Maryn McKenna, *Superbug: The Fatal Menace of MRSA* (New York: Simon and Schuster, 2010).

9 For basics and importance of efflux pumps, see M. A. Webber and L. J. V. Piddock, "The Importance of Efflux Pumps in Bacterial Antibiotic Resistance," *Journal of Antimicrobial Chemotherapy* 51, no. 1 (2003): 9–11.

10 Karen Bush, "Past and Present Perspectives on β-lactamases," *Antimicrobial Agents and Chemotherapy* 62, no. 10 (2018): e01076–18.

11 David E. Pettijohn, "Structure and Properties of the Bacterial Nucleoid," *Cell* 30, no. 3 (1982): 667–69.

12 For the potential global impact, see the fact sheet from the World Health Organization available at https://www.who.int/news-room/fact-sheets/detail/antibiotic-resistance.

CHAPTER 2: FIFTY MILLION DEAD

1 Influenza Archive, City of Boston, https://www.influenzaarchive.org/cities/city-boston.html#.

2 Laura Spinney, "Vital Statistics: How the Spanish Flu of 1918 Changed India," *Caravan*, October 19, 2018.

3 Amir Afkhami, "Compromised Constitutions: The Iranian Experience with the 1918 Influenza Pandemic," *Bulletin of the History of Medicine* 77, no. 2 (2003): 367–92.

4 Sandra M. Tomkins, "The Influenza Epidemic of 1918–19 in Western Samoa," *Journal of Pacific History* 27, no. 2 (1992): 181–97.

5   NIH News Report. August 19, 2008. "Tuesday, August 19, 2008, "Bacterial Pneumonia Caused Most Deaths in 1918 Influenza Pandemic," https://www.nih.gov/news-events/news-releases/bacterial-pneumonia -caused-most-deaths-1918-influenza-pandemic.

6   Fred Rosner, "The Life of Moses Maimonides, a Prominent Medieval Physician," *Einstein Quarterly Journal of Biology and Medicine* 19 (2002): 125–28.

7   Ibid.

8   Robert D. Purrington, *The First Professional Scientist: Robert Hooke and the Royal Society of London*, vol. 39 (New York: Springer Science & Business Media, 2009).

9   Howard Gest, "The Discovery of Microorganisms by Robert Hooke and Antoni Van Leeuwenhoek, Fellows of the Royal Society," *Notes and Records of the Royal Society of London* 58, no. 2 (2004): 187–201.

10 Jan van Zuylen, "The Microscopes of Antoni van Leeuwenhoek," *Journal of Microscopy* 121, no. 3 (1981): 309–28.

11 J. R. Porter, "Antony van Leeuwenhoek: Tercentenary of His Discovery of Bacteria," *Bacteriological Reviews* 40, no. 2 (1976): 260.

12 Nick Lane, "The Unseen World: Reflections on Leeuwenhoek (1677) 'Concerning Little Animals,'" *Philosophical Transactions of the Royal Society B: Biological Sciences* 370, no. 1666 (2015): 2014034.

13 F. H. Garrison, "Edwin Klebs (1834–1913)," *Science* 38, no. 991 (1913): 920–21.

14 For a detailed account of Sternberg's life, see Martha L. Sternberg, *George Miller Sternberg: A Biography* (Chicago: American Medical Association, 1920).

15 George Sternberg, "The Pneumonia-Coccus of Friedlander (*Micrococcus pasteuri*, Sternberg)," *American Journal of the Medical Sciences* 179 (1885): 106–22.

16 Leonard D. Epifano and Robert D. Brandstetter, "Historical Aspects of Pneumonia," in *The Pneumonias* (New York: Springer, 1993), 1–14.

17 Robert Austrian, "The Gram Stain and the Etiology of Lobar Pneumonia, an Historical Note," *Bacteriological Reviews* 24, no. 3 (1960): 261.

18 Ibid.

19 Ibid.

20 Ibid.

21 Carl Friedlaender, *The Use of the Microscope in Clinical and Pathological Examinations* (New York: D. Appleton, 1885), 75.

22 Ibid., 76.

23 Robert Austrian, "The Gram Stain and the Etiology of Lobar Pneumonia, an Historical Note," *Bacteriological Reviews* 24, no. 3 (1960): 261.

24 B. B. Biswas, P. S. Basu, and M. K. Pal, "Gram Staining and Its Molecular Mechanism," *International Review of Cytology* 29 (1970), 1–27.

CHAPTER 3: TIME AND SPACE

1   Richard J. White, "The Early History of Antibiotic Discovery: Empiricism Ruled," in *Antibiotic Discovery and Development* (Boston: Springer, 2012), 3–31.

2   Vanessa M. D'Costa, Katherine M. McGrann, Donald W. Hughes, and Gerard D. Wright, "Sampling the Antibiotic Resistome," *Science* 311, no. 5759 (2006): 374–77.

3   H. A. Barton, "Much Ado About Nothing: Cave Cultivar Collections," *Society for Industrial Microbiology Annual Meeting*, San Diego, CA, August 13, 2008.

4   For more information about the discovery of "Deep Secrets," see Stephen Reames, Lawrence Fish, Paul Burger, and Patricia Kambesis, *Deep Secrets: The Discovery and Exploration of Lechuguilla Cave* (St. Louis, MO: Cave Books, 1999), chaps. 1–2.

5   Shayla Love, "This Woman Is Exploring Deep Caves to Find Ancient Antibiotic Resistance," *Vice Magazine*, April 18, 2018.

6   E. Yong, "Isolated for Millions of Years, Cave Bacteria Resist Modern Antibiotics," *National Geographic*, April 13, 2012.

7   Kirandeep Bhullar, Nicholas Waglechner, Andrew Pawlowski, Kalinka Koteva, Eric D. Banks, Michael D. Johnston, Hazel A. Barton et al., "Antibiotic Resistance Is Prevalent in an Isolated Cave Microbiome," *PloS One* 7, no. 4 (2012): e34953.

8   Ibid.

9   Ibid.

10 Based on the author's interview with Gerry Wright, August 8, 2018.

CHAPTER 4: FRIENDS IN FAR PLACES

1   M. Fessenden, "Here's How Cinnamon Is Harvested in Indonesia," *Smithsonian*, April 22, 2015.

2   For a photo essay on the Yanomami, see "The Yanomami: An Isolated Yet Imperiled Amazon Tribe," *Washington Post*, July 25, 2014.

3   Amin Talebi Bezmin Abadi, "*Helicobacter pylori*: A Beneficial Gastric Pathogen?" *Frontiers in Medicine* 1 (2014): 26.

4   Chandrabali Ghose, Guillermo I. Perez-Perez, Maria-Gloria Dominguez-Bello, David T. Pride, Claudio M. Bravi, and Martin J. Blaser, "East Asian Genotypes of *Helicobacter pylori* Strains in Amerindians Provide Evidence for Its Ancient Human Carriage," *Proceedings of the National Academy of Sciences* 99, no. 23 (2002): 15107–111.

5   Maria Gloria Dominguez-Bello, "A Microbial Anthropologist in the Jungle," *Cell* 167, no. 3 (2016): 588–94.

6   CBC Radio, "Amazon Tribe's Gut Bacteria Reveals Toll of Western Lifestyle," April 20, 2015.

7   Jose C. Clemente, Erica C. Pehrsson, Martin J. Blaser, Kuldip Sandhu, Zhan Gao, Bin Wang, Magda Magris et al., "The Microbiome of Uncontacted Amerindians," *Science Advances* 1, no. 3 (2015): e1500183.

8   Ibid.

9   Based on the author's interview with Gautam Dantas, March 8, 2019.

CHAPTER 5: NEAR THE SEED VAULT

1   Based on the author's interview with Gerry Wright, August 8, 2018.

2   K. Crowe, "Antibiotic-Resistant Bacteria Disarmed with Fungus Compound," CBC News, June 25, 2014.

3   Ibid.

4   For more information about the Global Seed Vault, see https://www.seed-vault.no.

5   Clare M. McCann, Beate Christgen, Jennifer A. Roberts, Jian-Qiang Su, Kathryn E. Arnold, Neil D. Gray, Yong-Guan Zhu et al., "Understanding Drivers of Antibiotic Resistance Genes in High Arctic Soil Ecosystems," *Environment International* 125 (2019): 497–504.

CHAPTER 6: KEYS TO KARACHI

1   Tony Kirby, "Timothy Walsh: Introducing the World to NDM-1," *Lancet Infectious Diseases* 12, no. 3 (2012): 189.
2   Based on the author's interview with Timothy Walsh, September 4, 2018.
3   Ibid.
4   Patrice Nordmann, Laurent Poirel, Mark A. Toleman, and Timothy R. Walsh, "Does Broad-Spectrum Beta-lactam Resistance Due to NDM-1 Herald the End of the Antibiotic Era for Treatment of Infections Caused by Gram-Negative Bacteria?," *Journal of Antimicrobial Chemotherapy* 66, no. 4 (2011): 689–92.
5   Kate Kelland and Ben Hirschler, "Scientists Find New Superbug Spreading from India," Reuters, August 11, 2010.
6   Sarah Boseley, "Are You Ready for a World Without Antibiotics?," *Guardian*, August 12, 2010.
7   Geeta Pandey, "India Rejects UK Scientists' 'Superbug' Claim," BBC News, August 12, 2010.
8   J. Sood, *Superbug: India Gets Bugged. Government Downplays Threat from Drug-Resistant Bacteria*, available at https://www.downtoearth.org.in/news /superbug-india-gets-bugged-1850.
9   The controversy around the name continues to a certain extent to this date. For more details, see Timothy R. Walsh and Mark A. Toleman, "The New Medical Challenge: Why NDM-1? Why Indian?," *Expert Review of Anti-Infective Therapy* 9, no. 2 (2011): 137–41; and G. Nataraj, "New Delhi Metallo Beta-Lactamase: What Is in a Name?," *Journal of Postgraduate Medicine* 56, no. 4 (2010): 251; and "'New Delhi' Superbug Named Unfairly, Says *Lancet* Editor," BBC News, January 12, 2011.
10 Naomi Lubick, "Antibiotic Resistance Shows Up in India's Drinking Water," *Nature*, April 7, 2011.
11 "Travelers May Spread Drug-Resistant Gene from South Asia," VOA News, April 26, 2011, available at https://learningenglish.voanews.com/a /india-superbug-120747334/115208.html.
12 T. V. Padma, "India Questions 'Superbug' Conclusions, Research Ethics," SciDev.Net, April 8, 2011.

CHAPTER 7: WAR AND PEACE

1   Chung King-thom and Liu Jong-kang, *Pioneers in Microbiology: The Human Side of Science* (Singapore: World Scientific, 2017), 221–22.
2   Venita Jay, "The Legacy of Robert Koch," *Archives of Pathology & Laboratory Medicine* 125, no. 9 (2001): 1148–49.
3   Ibid.
4   William Rosen, *Miracle Cure: The Creation of Antibiotics and the Birth of Modern Medicine* (New York: Penguin, 2017), 22–23.
5   Ibid., 23–24.

6   Gerhart Drews, "Ferdinand Cohn, a Founder of Modern Microbiology," *ASM News* 65, no. 8 (1999): 54.

7   Lawrason Brown, "Robert Koch," *Bulletin of the New York Academy of Medicine* 8, no. 9 (1932): 558.

8   William Rosen, *Miracle Cure*, 25.

9   Steve M. Blevins and Michael S. Bronze, "Robert Koch and the 'Golden Age' of Bacteriology," *International Journal of Infectious Diseases* 14, no. 9 (2010): e744–51.

10 H. R. Wiedeman, "Robert Koch," *European Journal of Pediatrics* 149, no. 4 (1990): 223.

11 William Rosen, *Miracle Cure*, 28.

12 Florian Winau, Otto Westphal, and Rolf Winau, "Paul Ehrlich—in Search of the Magic Bullet," *Microbes and Infection* 6, no. 8 (2004): 786–89.

13 William Rosen, *Miracle Cure*, 39–49.

14 Gian Franco Gensini, Andrea Alberto Conti, and Donatella Lippi, "The Contributions of Paul Ehrlich to Infectious Disease," *Journal of Infection* 54, no. 3 (2007): 221–24.

15 Hiroshi Maruta, "From Chemotherapy to Signal Therapy (1909–2009): A Century Pioneered by Paul Ehrlich," *Drug Discoveries & Therapeutics* 3, no. 2 (2009).

16 William Rosen, *Miracle Cure*, 55.

17 Stefan H. E. Kaufmann, "Paul Ehrlich: Founder of Chemotherapy," *Nature Reviews Drug Discovery* 7, no. 5 (2008): 373.

18 B. Lee Ligon, "Robert Koch: Nobel Laureate and Controversial Figure in Tuberculin Research," in *Seminars in Pediatric Infectious Diseases* 13, no. 4 (2002): 289–99.

19 William Rosen, *Miracle Cure*, 30.

20 Wolfgang U. Eckart, "The Colony as Laboratory: German Sleeping Sickness Campaigns in German East Africa and in Togo, 1900–1914," *History and Philosophy of the Life Sciences* 24, no. 1 (February 2002): 69–89.

21 Ibid.

22 Ibid.

23 John Lichfield, "De Gaulle Named Greatest Frenchman in Television Poll," *Independent*, April 6, 2005, https://www.independent.co.uk/news/world /europe/de-gaulle-named-greatest-frenchman-in-television-poll-531330 .html.

24 William Rosen, *Miracle Cure*, 16–20.

25 Gerald L. Geison, "Organization, Products, and Marketing in Pasteur's Scientific Enterprise," *History and Philosophy of the Life Sciences* 24, no. 1 (2002): 37–51.

26 William Rosen, *Miracle Cure*, 26–28.

27 Maxime Schwartz, "Louis Pasteur and Molecular Medicine: A Centennial Celebration," *Molecular Medicine* 1 (September 1995): 593.

28 L. Robbins, *Louis Pasteur and the Hidden World of Microbes* (New York: Oxford University Press, 2001).

29 Gerald L. Geison, *The Private Science of Louis Pasteur*, vol. 306 (Princeton, NJ: Princeton University Press, 2014).

30 Ibid.

31 Julie Ann Miller, "The Truth About Louis Pasteur," *BioScience* 43, no. 5 (1993): 280–82.

CHAPTER 8: FROM THE PHAGES OF HISTORY

1 Martha R. J. Clokie, Andrew D. Millard, Andrey V. Letarov, and Shaun Heaphy, "Phages in Nature," *Bacteriophage* 1, no. 1 (2011): 31–45.

2 Félix d'Hérelle and George H. Smith, *The Bacteriophage and Its Behavior* (Baltimore: Williams & Wilkins, 1926).

3 Donna H. Duckworth and Paul A. Gulig, "Bacteriophages," *BioDrugs* 16, no. 1 (2002): 57–62.

4 "On an Invisible Microbe Antagonistic to Dysentery Bacilli," note by M. F. d'Hérelle, presented by M. Roux, *Comptes Rendus Academie des Sciences* 1917, 165:373–5; *Bacteriophage* 1, no. 1 (2011), 3–5, DOI: 10.4161/bact.1.1.14941.

5 Félix H. d'Hérelle, *Le Bacteriophage*, vol. 5 (Paris: Masson, 1921).

6 William C. Summers, "The Strange History of Phage Therapy," *Bacteriophage* 2, no. 2 (2012): 130–33.

7 Donna H. Duckworth, "Who Discovered Bacteriophage?" *Bacteriological Reviews* 40, no. 4 (1976): 793.

8 Antony Twort, *In Focus, Out of Step: A Biography of Frederick William Twort F.R.S., 1877–1950* (Gloucestershire, UK: Sutton Pub. Ltd., 1993).

9 F. W. Twort, "An Investigation on the Nature of Ultra-Microscopic Viruses," *Lancet* 186, no. 4814 (December 1915): 1241–43.

10 Donna H. Duckworth, "Who Discovered Bacteriophage?," 793.

11 Paul Gordon Fildes, "Frederick William Twort, 1877–1950," obituary, *Royal Society* (1951): 505–17.

12 William C. Summers, "The Strange History of Phage Therapy," 130–33.

13 Félix d'Herelle, Reginald Hampstead Malone, and Mahendra Nath Lahiri, *Studies on Asiatic Cholera, Indian Medical Research Memoirs, no.* 14 (Calcutta: Pub. for the Indian Research Fund Association by Thacker, Spink & Co., 1930).

14 William C. Summers, "On the Origins of the Science in *Arrowsmith*: Paul de Kruif, Félix d'Hérelle, and Phage," *Journal of the History of Medicine and Allied Sciences* 46, no. 3 (1991): 315–32.

15 Ernst W. Caspari and Robert E. Marshak, "The Rise and Fall of Lysenko," *Science* 149, no. 3681 (1965): 275–78.

16 Richard Stone, "Stalin's Forgotten Cure," *Science* 298 (2002): 728–31.

17 Dmitriy Myelnikov, "An Alternative Cure: The Adoption and Survival of Bacteriophage Therapy in the USSR, 1922–1955," *Journal of the History of Medicine and Allied Sciences* 73, no. 4 (2018): 385–411.

18 Ibid.

19 Anna Kuchment, *The Forgotten Cure: The Past and Future of Phage Therapy* (New York: Springer Science & Business Media, 2011), 26–34.

20 Ibid.

21 Richard Stone, "Stalin's Forgotten Cure," 728–31.

CHAPTER 9: SULFA AND THE WAR

1 Copy of the letter available at https://teslauniverse.com/nikola-tesla/letters/june-12th-1931-letter-waldemar-kaempffert-nikola-tesla.

2 Waldemar Kaempffert, "News of Dr. Paul Gelmo, Discoverer of Sulfanil-amide," *Journal of the History of Medicine and Allied Sciences* 5 (Spring 1950): 213–14.

3 William Rosen, *Miracle Cure: The Creation of Antibiotics and the Birth of Modern Medicine* (New York: Penguin, 2017), 70.

4 C. Jeśman, A. Młudzik, and M. Cybulska, "History of Antibiotics and Sul-phonamides Discoveries," *Polski Merkuriusz Lekarski: Organ Polskiego Towarzystwa Lekarskiego* 30, no. 179 (2011): 320–32.

5 Matt McCarthy, *Superbugs: The Race to Stop an Epidemic* (New York: Avery /Penguin, 2019), 29–37.

6 Mark Wainwright and Jette E. Kristiansen, "On the 75th Anniversary of Prontosil," *Dyes and Pigments* 88, no. 3 (2011): 231–34.

7 Ibid.

8 William Rosen, *Miracle Cure*, 70.

9 M. Spring, "A Brief Survey of the History of the Antimicrobial Agents," *Bulletin of the New York Academy of Medicine* 51, no. 9 (1975): 101.

10 Carol Ballentine, "Taste of Raspberries, Taste of Death: The 1937 Elixir Sulfanilamide Incident," *FDA Consumer Magazine* 15, no. 5 (1981).

11 Ibid.

12 Paul M. Wax, "Elixirs, Diluents, and the Passage of the 1938 Federal Food, Drug and Cosmetic Act," *Annals of Internal Medicine* 122, no. 6 (1995): 456–61.

13 Muhammad H. Zaman, *Bitter Pills: The Global War on Counterfeit Drugs* (New York: Oxford University Press, 2018), 64–94.

14 Dr. Elliott Cutler's papers and letters are in the archives at the Countway Library of Medicine at Harvard University.

15 Arnold Lorentz Ahnfeldt, Robert S. Anderson, John Boyd Coates, Calvin H. Goddard, and William S. Mullins, *The Medical Department of the United States Army in World War II*, vol. 2 (Washington, DC: Office of the Surgeon General, Department of the Army, 1964), 67.

16 Ibid.

17 Ibid.

18 Dr. Elliott Cutler's papers and letters are in the archives at the Countway Library of Medicine at Harvard University.

CHAPTER 10: MOLD JUICE

1 Nobel lecture by Dr. Fleming, December 11, 1945, available at https://www.nobelprize.org/uploads/2018/06/fleming-lecture.pdf.

2 Ibid.

3 Kevin Brown, *Penicillin Man: Alexander Fleming and the Antibiotic Revolution* (Cheltenham, UK: History Press, 2005).

4 Ronald Hare, *The Birth of Penicillin, and the Disarming of Microbes* (Crows Nest, Australia: George Allen and Unwin, 1970).

5 William Rosen, *Miracle Cure: The Creation of Antibiotics and the Birth of Modern Medicine* (New York: Penguin, 2017), 70.

6 Joan W. Bennett and King-Thom Chung, "Alexander Fleming and the Discovery of Penicillin," *Advances in Applied Microbiology* 49 (2001): 163–84.

7 Robert Bud, *Penicillin: Triumph and Tragedy* (Peterborough, UK: Oxford University Press on Demand, 2007), 23–32.

8   Joan W. Bennett and King-Thom Chung, "Alexander Fleming," 163–84.

9   William Rosen, *Miracle Cure*, 103–15.

10 Carol L. Moberg, "Penicillin's Forgotten Man: Norman Heatley; Although He's Been Overlooked, His Skills in Growing Penicillin Were a Key to Florey and Chain's Clinical Trials," *Science* 253, no. 5021 (1991): 734–36.

11 Ibid.

12 Ibid.

13 Robert Bud, *Penicillin*, 30–40.

14 William Rosen, *Miracle Cure*, 127.

15 Robert Bud, *Penicillin*, 32–33.

16 Eric Lax, *The Mold in Dr. Florey's Coat: The Story of the Penicillin Miracle* (New York: Macmillan, 2004): 170–73.

17 William Rosen, *Miracle Cure*, 133.

18 Eric Lax, *The Mold in Dr. Florey's Coat*, 204–23.

19 Ibid., 186.

20 Alfred N. Richards, "Production of Penicillin in the United States (1941–1946)," *Nature* 201, no. 4918 (1964): 441–45.

21 Eric Lax, *The Mold in Dr. Florey's Coat*, 185–89.

22 William Rosen, *Miracle Cure*, 138–41.

23 Ibid.

CHAPTER 11: TABLETS FROM TEARS

1   Letters to Dr. Ermolieva and other Soviet scientists, as part of the invitation to join expert committee on antibiotics, are available in the WHO archive.

2   Ibid.

3   For detailed biography of Z. Ermolieva, see S. Navashin, "Obituary: Prof. Zinaida Vissarionouna Ermolieva," *Journal of Antibiotics* 28, no. 5 (1975): 399; see also the work of Anna Eremeeva on the early life of Z. Ermolieva.

4   Stuart Mudd, "Recent Observations on Programs for Medicine and National Health in the USSR," *Proceedings of the American Philosophical Society* 91, no. 2 (1947): 181–88.

5   Ibid.

6   Anna Kuchment, "'They're Not a Panacea': Phage Therapy in the Soviet Union and Georgia," in *The Forgotten Cure* (New York: Springer, 2012), 53–62.

7   Dmitriy Myelnikov, "An Alternative Cure: The Adoption and Survival of Bacteriophage Therapy in the USSR, 1922–1955," *Journal of the History of Medicine and Allied Sciences* 73, no. 4 (2018): 385–411.

8   Lev L. Kisselev, Gary I. Abelev, and Feodor Kisseljov, "Lev Zilber, the Personality and the Scientist," in *Advances in Cancer Research*, vol. 59 (Cambridge, MA: Academic Press, 1992), 1–40.

9   Dmitriy Myelnikov, "An Alternative Cure," 385–411.

CHAPTER 12: THE NEW PANDEMIC

1   Robert Bud, *Penicillin: Triumph and Tragedy* (Peterborough, UK: Oxford University Press on Demand, 2007), 118–19.

2   Mary Barber, "Staphylococcal Infection Due to Penicillin-Resistant Strains," *British Medical Journal* 2, no. 4534 (1947): 863.

3 Ibid.

4 H. J. Bensted, "Central Public Health Laboratory, Colindale: New Laboratory Block," *Nature* 171, no. 4345 (1953): 248–49.

5 For details about the Public Health Laboratories, see UK National Archives on PHLS.

6 Ibid.

7 Kathryn Hillier, "Babies and Bacteria: Phage Typing, Bacteriologists, and the Birth of Infection Control," *Bulletin of the History of Medicine* 80, no. 4 (Winter 2006): 733–61.

8 Historical note of the University of Melbourne on Dr. Rountree, available at http://www.austehc.unimelb.edu.au/guides/roun/histnote.htm.

9 Ibid.

10 Kathryn Hillier, "Babies and Bacteria," 733–61.

11 Online museum of the University of Sydney, available at https://sydney.edu.au/medicine/museum/mwmuseum/index.php/Isbister,_Jean_Sinclair.

12 Kathryn Hillier, "Babies and Bacteria," 733–61.

13 Ibid.

14 Ibid.

15 For a discussion on post-penicillin antibiotics, particularly methicillin, see E. M. Tansey, ed. *Post Penicillin Antibiotics: From Acceptance to Resistance? A Witness Seminar*, Held at the Wellcome Institute for the History of Medicine, London, May 12, 1998 (London: Wellcome Trust, 2000).

16 Robert C. Moellering Jr., "MRSA: The First Half Century," *Journal of Antimicrobial Chemotherapy* 67, no. 1 (2011): 4–11.

17 Correspondence of Patricia Jevons in *British Medical Journal*, January 14, 1961, available at https://www.ncbi.nlm.nih.gov/pmc/articles/PMC1952878/pdf/brmedj02876-0103.pdf.

18 BBC News report on World War II, available at https://www.bbc.co.uk/history/ww2peopleswar/stories/15/a2099315.shtml.

19 E. M. Tansey, ed., *Post Penicillin Antibiotics*.

20 Fred F. Barrett, Read F. McGehee Jr., and Maxwell Finland, "Methicillin-Resistant *Staphylococcus aureus* at Boston City Hospital: Bacteriologic and Epidemiologic Observations," *New England Journal of Medicine* 279, no. 9 (1968): 441–48.

CHAPTER 13: THE MAN IN THE BLUE MUSTANG

1 Based on the author's interview with Dr. Ron Arky, February 14, 2019.

2 For biographical notes on Dr. Finland, see those available in the US National Academy of Sciences. See also Jerome O. Klein, Carol J. Baker, Fred Barrett, and James D. Cherry, "Maxwell Finland, 1902–1987: A Remembrance," *Pediatric Infectious Disease Journal* 21, no. 3 (2002): 181; Jerome O. Klein, "Maxwell Finland: A Remembrance," *Clinical Infectious Diseases* 34, no. 6 (March 2002): 725–29; as well as his obituaries published in various newspapers.

3 Arthur R. Reynolds, "Pneumonia: The New 'Captain of the Men of Death': Its Increasing Prevalence and the Necessity of Methods for Its Restriction," *Journal of the American Medical Association* 40, no. 9 (1903): 583–86.

4   Harry M. Marks, *The Progress of Experiment: Science and Therapeutic Reform in the United States, 1900–1990* (Cambridge: Cambridge University Press, 2000), 106–7.

5   The letters and minutes of various meetings where Finland and Waksman express their strong positions are available at the World Health Organization archives in Geneva.

6   Scott H. Podolsky, "To Finland and Back," *Harvard Medicine Magazine*, summer 2013.

7   For more details about the scandal see Richard E. McFadyen, "The FDA's Regulation and Control of Antibiotics in the 1950s: The Henry Welch Scandal, Félix Martí-Ibáñez, and Charles Pfizer & Co.," *Bulletin of the History of Medicine* 53, no. 2 (1979): 159–69.

8   Ibid.

9   Scott H. Podolsky, *The Antibiotic Era: Reform, Resistance, and the Pursuit of a Rational Therapeutics* (Baltimore: Johns Hopkins University Press, 2015).

CHAPTER 14: HONEYMOON

1   Charles Drechsler, "Morphology of the Genus Actinomyces. I," *Botanical Gazette* 67, no. 1 (1919): 65–83.

2   Antonio H. Romano and Robert S. Safferman, "Studies on Actinomycetes and Their Odors," *Journal of the American Water Works Association* 55, no. 2 (1963): 169–76.

3   R. G. Benedict, "Antibiotics Produced by Actinomycetes," *Botanical Review* 19, no. 5 (1953): 229.

4   John Simmons, *Doctors and Discoveries: Lives That Created Today's Medicine* (New York: Houghton Mifflin Harcourt, 2002), 259.

5   Thomas M. Daniel, *Pioneers of Medicine and Their Impact on Tuberculosis* (Rochester, NY: University of Rochester Press, 2000), 180.

6   Peter Pringle, *Experiment Eleven: Dark Secrets Behind the Discovery of a Wonder Drug* (London: Bloomsbury Publishing, 2012), 27–60.

7   For a detailed account of the Schatz-Waksman affair, see Peter Pringle, *Experiment Eleven*.

8   William Rosen, *Miracle Cure: The Creation of Antibiotics and the Birth of Modern Medicine* (New York: Penguin, 2017), 203–6.

9   Ibid.

10 Wolfgang Minas, "Erythromycins," *Encyclopedia of Industrial Biotechnology: Bioprocess, Bioseparation, and Cell Technology* (2009): 1–14.

11 Johanna Son, "Who Really Discovered Erythromycin?," IPS News. November 9, 1994.

12 Obituary of Reverend Bouw, *Toledo Blade*, July 4, 2006.

13 D. J. McGraw, *The Antibiotic Discovery Era (1940–1960): Vancomycin as an Example of the Era*, PhD dissertation, 1974, Oregon State University, Corvallis, OR.

14 Donald P. Levine, "Vancomycin: A History," *Clinical Infectious Diseases* 42, no. S1 (2006): S5–S12.

15 Michael White, "Elizabeth Taylor, My Great-Grandpa, and the Future of Antibiotics," *Pacific Standard*, January 22, 2015.

16 E. M. Tansey, ed., *Post Penicillin Antibiotics: From Acceptance to Resistance?*,

A Witness Seminar, held at the Wellcome Institute for the History of Medicine, London, May 12, 1998 (London: Wellcome Trust, 2000).

17 Kristine Krafts, Ernst Hempelmann, and Agnieszka Skórska-Stania, "From Methylene Blue to Chloroquine: A Brief Review of the Development of an Antimalarial Therapy," *Parasitology Research* 111, no. 1 (2012): 1–6.

18 D. J. Wallace, "The History of Antimalarials," *Lupus* 5, no. S1 (1996): S2–3.

19 Claude Mazuel, "Norfloxacin," in *Analytical Profiles of Drug Substances*, vol. 20 (Cambridge, MA: Academic Press, 1991), 557–600.

20 Hisashi Takahashi, Isao Hayakawa, and Takeshi Akimoto, "The History of the Development and Changes of Quinolone Antibacterial Agents," *Yakushigaku Zasshi* 38, no. 2 (2003): 161–79.

21 Vincent T. Andriole, "The Future of the Quinolones," *Drugs* 45, no. 3 (1993): 1–7.

22 Dan Prochi, "Bayer's $74M Pay-for-Delay Deal Approved in Calif.," Law360, November 18, 2013.

23 T. E. Daum, D. R. Schaberg, M. S. Terpenning, W. S. Sottile, and C. A. Kauffman, "Increasing Resistance of *Staphylococcus aureus* to Ciprofloxacin," *Antimicrobial Agents and Chemotherapy* 34, no. 9 (1990): 1862–63.

CHAPTER 15: MATING BACTERIA

1 Obituary of Dr. Joshua Lederberg, *Guardian*, February 11, 2008.

2 Stephen S. Morse, "Joshua Lederberg (1925–2008)," *Science* 319, no. 5868 (2008): 1351.

3 Miriam Barlow, "What Antimicrobial Resistance Has Taught Us About Horizontal Gene Transfer," in *Horizontal Gene Transfer* (Totowa, NJ: Humana Press, 2009), 397–411.

4 M. L. Morse, Esther M. Lederberg, and Joshua Lederberg, "Transduction in *Escherichia coli* K-12," *Genetics* 41, no. 1 (1956): 14.

5 Autobiographical notes and memoir of Professor Toshio Fukasawa shared with the author.

6 See Tsutomu Watanabe, "Infectious Drug Resistance in Enteric Bacteria," *New England Journal of Medicine* 275, no. 16 (1966): 888–94, and Tsutomu Watanabe, "Infective Heredity of Multiple Drug Resistance in Bacteria," *Bacteriological Reviews* 27, no. 1 (1963): 87.

CHAPTER 16: S IS FOR SOVIET

1 For a detailed account of Lysenko's impact on Soviet genetics, see Peter Pringle, *The Murder of Nikolai Vavilov* (New York: Simon and Schuster, 2008). See also Simon Ings, *Stalin and the Scientists* (Boston: Atlantic Monthly Press, 2017), and Loren Graham, *Lysenko's Ghost* (Cambridge, MA: Harvard University Press, 2016).

2 Valery N. Soyfer, "New Light on the Lysenko Era," *Nature* 339, no. 6224 (1989): 415.

3 Yasha M. Gall and Mikhail B. Konashev, "The Discovery of Gramicidin S: The Intellectual Transformation of GF Gause from Biologist to Researcher of Antibiotics and on Its Meaning for the Fate of Russian Genetics," *History and Philosophy of the Life Sciences* (2001): 137–50.

4 For a biographical note on the life of Gause, see Nikolai N. Vorontsov and

Jakov M. Gall, "Georgyi Frantsevich Gause 1910–1986," *Nature* 323, no. 6084 (1986): 113; and J. M. Gall, *Georgi Franzevich Gause* (St. Petersburg: Nestor-Historia, 2012), in Russian.

5   Based on the author's interview with Gause's biographer, Yakov Gall, December 11, 2018.

6   Based on the author's interview with Dr. Wolfgang Witte, conducted in Berlin on June 25, 2018.

7   Mark C. Enright, D. Ashley Robinson, Gaynor Randle, Edward J. Feil, Hajo Grundmann, and Brian G. Spratt, "The Evolutionary History of Methicillin-Resistant *Staphylococcus aureus* (MRSA)," *Proceedings of the National Academy of Sciences* 99, no. 11 (2002): 7687–92.

8   Hartmut Berghoff and Uta Andrea Balbier, eds. *The East German Economy, 1945–2010: Falling Behind or Catching Up?* (Cambridge, UK: Cambridge University Press), 2013.

CHAPTER 17: THE NAVY BOYS

1   Based on the author's interviews with Dr. King K. Holmes, August 12, 2018, and September 26, 2019.

2   Peter J. Rimmer, "US Western Pacific Geostrategy: Subic Bay Before and After Withdrawal," *Marine Policy* 21, no. 4 (1997): 325–44.

3   Gerald R. Anderson, *Subic Bay from Magellan to Pinatubo: The History of the US Naval Station, Subic Bay* (Scotts Valley, CA: CreateSpace, 2009).

4   King K. Holmes, David W. Johnson, Thomas M. Floyd, and Paul A. Kvale, "Studies of Venereal Disease II: Observations on the Incidence, Etiology, and Treatment of the Postgonococcal Urethritis Syndrome," *Journal of the American Medical Association* 202, no. 6 (1967): 467–47.

5   King K. Holmes, David W. Johnson, and Thomas M. Floyd, "Studies of Venereal Disease I: Probenecid-Procaine Penicillin G Combination and Tetracycline Hydrochloride in the Treatment of Penicillin-Resistant Gonorrhea in Men," *Journal of the American Medical Association* 202, no. 6 (1967): 461–66.

6   Ibid.

7   Mari Rose Aplasca de los Reyes, Virginia Pato-Mesola, Jeffrey D. Klausner, Ricardo Manalastas, Teodora Wi, Carmelita U. Tuazon, Gina Dallabetta, et al., "A Randomized Trial of Ciprofloxacin Versus Cefixime for Treatment of Gonorrhea After Rapid Emergence of Gonococcal Ciprofloxacin Resistance in the Philippines," *Clinical Infectious Diseases* 32, no. 9 (2001): 1313–18.

CHAPTER 18: FROM ANIMALS TO HUMANS

1   Hugh Pennington, "Our Ability to Cope with Food Poisoning Outbreaks Has Not Improved Much in 50 Years," The Conversation, May 6, 2014.

2   Ibid.

3   Jim Phillips, David F. Smith, H. Lesley Diack, T. Hugh Pennington, and Elizabeth M. Russell, *Food Poisoning, Policy and Politics: Corned Beef and Typhoid in Britain in the 1960s* (Woodbridge, UK: Boydell Press, 2005), xiv, 334.

4   Robert Bud, *Penicillin: Triumph and Tragedy* (Peterborough, UK: Oxford University Press on Demand, 2007), 176–77.

5   Ibid.

6 Ibid., 178–79.

7 Ibid., 180–81.

8 Ibid., 182–84.

9 Mary D. Barton, "Antibiotic Use in Animal Feed and Its Impact on Human Health," *Nutrition Research Reviews* 13, no. 2 (2000): 279–99.

10 Claas Kirchhelle, "Swann Song: Antibiotic Regulation in British Livestock Production (1953–2006)," *Bulletin of the History of Medicine* 92, no. 2 (2018): 317–50.

11 Leading among the scientists against any cuts to antibiotics was Thomas Jukes. After the Swann report in the UK, when the FDA started to look at farms in the United States and had made recommendations about curtailing antibiotics in farm animals, Jukes was outraged. In 1970, he wrote in *The New England Journal of Medicine* that "the use of antibiotics for farm animals does not present a hazard to public health," and in 1977 he called the entire debate on food additives bizarre. He had called FDA scientists quacks before in his writing as well. In the same *New England Journal of Medicine* he wrote, "The most injurious of all 'food additives' is the additional food that is eaten after caloric needs have been satisfied. Overconsumption of food leads to obesity, which is a far greater danger to health than any of the food additives whose safety is now being questioned." Jukes was an unrepentant supporter of using modern chemicals for human welfare. Jukes would use his activism and forceful voice not just for antibiotics but also to argue against the ban on DDT in the light of a growing environmental movement in California and elsewhere.

12 Stuart Levy died in 2019. For his obituary, see Harrison Smith, *Washington Post*, September 19, 2019, available at https://www.washingtonpost .com/local/obituaries/stuart-levy-microbiologist-who-sounded-alarm -on-antibiotic-resistance-dies-at-80/2019/09/19/4011ea96-dae9–11e9-a688 –303693fb4b0b_story.html.

13 Maryn McKenna, *Big Chicken: The Incredible Story of How Antibiotics Created Modern Agriculture and Changed the Way the World Eats* (Boone, IA: National Geographic Books, 2017), 110–18.

14 McDonald's announced in 2003 that it would "requir[e] its meat suppliers to stop using antibiotics important in human medicine to promote animal growth based on findings from APUA's FAAIR report as scientific evidence"; see https://apua.org/ourhistory.

CHAPTER 19: THE NORWEGIAN SALMON

1 Based on the author's interview with Tore Midtvedt in his Oslo home, January 13, 2018.

2 MIC stands for minimum inhibitory concentration. To identify minimum inhibitory concentration (MIC)—the minimum amount of drug required to kill the bacteria—lab researchers have used tubes filled with antibiotics in varying concentrations, then added bacteria to them, a method that dates back to Fleming in the 1920s. Some improvements were made between the 1920s and 1950s, but largely the method has remained the same; the only difference is that instead of large test tubes, researchers now use small microplates and put them in incubators to create conditions where

bacteria can grow. Some plates come already prepared; others require the researchers to use additives. Tens of different antibiotics, or different concentrations of antibiotics, can be used simultaneously. More recently there have been new, quantitative, and automated methods developed by various biotech companies.

3   Norway is the largest producer of salmon in the world. According to Food and Agriculture, Norway produces more than 1.233 million metric tons of salmon every year.

4   Decades later, the movie was made available again. It is now available at https://tv.nrk.no/serie/fisk-i-fangenskap/1988/FSFJ00000488.

5   Ingunn Sommerset, Bjørn Krossøy, Eirik Biering, and Petter Frost, "Vaccines for Fish in Aquaculture," *Expert Review of Vaccines* 4, no. 1 (2005): 89–101.

6   Information available on the website of the Norwegian Royal Family, https://www.kongehuset.no/nyhet.html?tid=165449&sek=26939.

CHAPTER 20: CLOSER TO SYDNEY THAN TO PERTH

1   Based on the author's interview with Professor Warren Grubb, February 1, 2019, and on autobiographical notes Professor Grubb kindly shared with the author.

2   See papers by Dr. Gracey, including Michael Gracey and Malcolm King, "Indigenous Health Part 1: Determinants and Disease Patterns," *Lancet* 374, no. 9683 (2009): 65–75; and Malcolm King, Alexandra Smith, and Michael Gracey, "Indigenous Health Part 2: The Underlying Causes of the Health Gap," *Lancet* 374, no. 9683 (2009): 76–85.

3   Sheryl Persson, *The Royal Flying Doctor Service of Australia: Pioneering Commitment, Courage and Success*, readhowyouwant.com, 2010.

4   Keiko Okuma, Kozue Iwakawa, John D. Turnidge, Warren B. Grubb, Jan M. Bell, Frances G. O'Brien, Geoffrey W. Coombs et al., "Dissemination of New Methicillin-Resistant *Staphylococcus aureus* Clones in the Community," *Journal of Clinical Microbiology* 40, no. 11 (2002): 4289–94.

CHAPTER 21: A CLASSLESS PROBLEM

1   Muhammad H. Zaman, *Bitter Pills: The Global War on Counterfeit Drugs* (Oxford: Oxford University Press, 2018).

2   Dinar Kale and Steve Little, "From Imitation to Innovation: The Evolution of R&D Capabilities and Learning Processes in the Indian Pharmaceutical Industry," *Technology Analysis & Strategic Management* 19, no. 5 (2007): 589–609.

3   Ibid.

4   Stefan Ecks, "Global Pharmaceutical Markets and Corporate Citizenship: The Case of Novartis' Anti-Cancer Drug Glivec," *BioSocieties* 3, no. 2 (2008): 165–81.

5   Muhammad H. Zaman, *Bitter Pills*.

6   Fatime Sheikh, "A France We Must Visit," *Friday Times*, June 15, 2018.

7   J. I. Tribunal, Batch J-093, "The Pathology of Negligence: Report of the Judicial Inquiry Tribunal to Determine the Causes of Deaths of Patients of the Punjab Institute of Cardiology, Lahore in 2011–2012" (2012).

8   "Nothing Wrong with Tyno Cough Syrup, Victims Overdosed," *Express Tribune*, November 27, 2012.
9   Muhammad H. Zaman, *Bitter Pills*, 26–50.
10 Sachiko Ozawa, Daniel R. Evans, Sophia Bessias, Deson G. Haynie, Tatenda T. Yemeke, Sarah K. Laing, and James E. Herrington, "Prevalence and Estimated Economic Burden of Substandard and Falsified Medicines in Low-and Middle-Income Countries: A Systematic Review and Meta-Analysis," *JAMA Network Open* 1, no. 4 (2018): e181662.
11 Muhammad H. Zaman, *Bitter Pills*, 60–75.
12 P. Sensi, "History of the Development of Rifampin," *Reviews of Infectious Diseases* 5, no. S3 (1983): S402–6.
13 Zohar B. Weinstein and Muhammad H. Zaman, "Evolution of Rifampin Resistance in *Escherichia coli* and *Mycobacterium smegmatis* Due to Substandard drugs," *Antimicrobial Agents and Chemotherapy* 63, no. 1 (2019): e01243–18.

CHAPTER 22: THE STUBBORN WOUNDS OF WAR

1   Aoife Howard, Michael O'Donoghue, Audrey Feeney, and Roy D. Sleator, "*Acinetobacter baumannii*: An Emerging Opportunistic Pathogen," *Virulence* 3, no. 3 (2012): 243–50.
2   Lenie Dijkshoorn, Alexandr Nemec, and Harald Seifert, "An Increasing Threat in Hospitals: Multidrug-Resistant *Acinetobacter baumannii*," *Nature Reviews Microbiology* 5, no. 12 (2007): 939.
3   Centers for Disease Control and Prevention (CDC), "*Acinetobacter baumannii* Infections Among Patients at Military Medical Facilities Treating Injured US Service Members, 2002–2004," *Morbidity and Mortality Weekly Report* 53, no. 45 (2004): 1063.
4   Pew Trusts Reports, "The Threat of Multidrug-Resistant Infections to the U.S. Military," March 1, 2012.
5   Rachel Nugent, "Center for Global Development Report," June 14, 2010.
6   Z. T. Sahli, A. R. Bizri, and G. S. Abu-Sittah, "Microbiology and Risk Factors Associated with War-Related Wound Infections in the Middle East," *Epidemiology & Infection* 144, no. 13 (2016): 2848–57.
7   Based on the author's several interviews with Dr. Abu Sittah, between August and October 2018.
8   Based on the author's interview with Dr. Souha Kanj, September 25, 2018.
9   Omar Dewachi, *Ungovernable Life: Mandatory Medicine and Statecraft in Iraq* (Palo Alto, CA: Stanford University Press, 2017).
10 Based on the author's interview with Dr. Omar Dewachi, October 2018.
11 Omar Dewachi, *Ungovernable Life*.
12 Based on the author's interview with Dr. Vinh-Kim Nguyen, March 21, 2019.
13 Prashant K. Dhakephalkar and Balu A. Chopade, "High Levels of Multiple Metal Resistance and Its Correlation to Antibiotic Resistance in Environmental Isolates of Acinetobacter," *Biometals* 7, no. 1 (1994): 67–74.

CHAPTER 23: COUNTING THE DEAD

1   Based on the author's interviews with Dr. Ramanan Laxminarayan, July 16, 2018, and March 4, 2019.

2   Michael Bennett's book about his father, *My Father: An American Story of Courage, Shattered Dreams, and Enduring Love*, published in 2012, talks in detail about his father's medical ordeals.

3   Eili Klein, David L. Smith, and Ramanan Laxminarayan, "Hospitalizations and Deaths Caused by Methicillin-Resistant *Staphylococcus aureus*, United States, 1999–2005," *Emerging Infectious Diseases* 13, no. 12 (2007): 1840.

4   Ramanan Laxminarayan, Precious Matsoso, Suraj Pant, Charles Brower, John-Arne Røttingen, Keith Klugman, and Sally Davies, "Access to Effective Antimicrobials: A Worldwide Challenge," *Lancet* 387, no. 10014 (2016): 168–75.

5   Pamela Das and Richard Horton, "Antibiotics: Achieving the Balance Between Access and Excess," *Lancet* 387, no. 10014 (2016): 102–4.

6   Ganesan Gowrisankar, Ramachandran Chelliah, Sudha Rani Ramakrishnan, Vetrimurugan Elumalai, Saravanan Dhanamadhavan, Karthikeyan Brindha, Usha Antony et al., "Chemical, Microbial and Antibiotic Susceptibility Analyses of Groundwater After a Major Flood Event in Chennai," *Scientific Data* 4 (2017): 170135.

CHAPTER 24: CLUES IN THE SEWAGE

1   Based on the author's interview with Dr. Frank Møller Aarestrup, April 2, 2019.

2   Thomas Nordahl Petersen, Simon Rasmussen, Henrik Hasman, Christian Carøe, Jacob Bælum, Anna Charlotte Schultz, Lasse Bergmark et al., "Meta-Genomic Analysis of Toilet Waste from Long Distance Flights; a Step Towards Global Surveillance of Infectious Diseases and Antimicrobial Resistance," *Scientific Reports* 5 (2015): 11444.

3   Ibid.

4   For more information about the project, see https://www.compare-europe.eu/Library/Global-Sewage-Surveillance-Project.

5   Rene S. Hendriksen, Patrick Munk, Patrick Njage, Bram Van Bunnik, Luke McNally, Oksana Lukjancenko, Timo Röder et al., "Global Monitoring of Antimicrobial Resistance Based on Metagenomics Analyses of Urban Sewage," *Nature Communications* 10, no. 1 (2019): 1124.

CHAPTER 25: X IS FOR EXTENSIVE

1   Based on the author's interview with Dr. Rumina Hasan, April 24, 2018.

2   Jeffrey D. Stanaway, Robert C. Reiner, Brigette F. Blacker, Ellen M. Goldberg, Ibrahim A. Khalil, Christopher E. Troeger, Jason R. Andrews et al., "The Global Burden of Typhoid and Paratyphoid Fevers: A Systematic Analysis for the Global Burden of Disease Study 2017," *Lancet Infectious Diseases* 19, no. 4 (2019): 369–81.

3   Zoe A. Dyson, Elizabeth J. Klemm, Sophie Palmer, and Gordon Dougan, "Antibiotic Resistance and Typhoid," *Clinical Infectious Diseases* 68, no. S2 (2019): S165–70.

4   Elizabeth J. Klemm, Sadia Shakoor, Andrew J. Page, Farah Naz Qamar, Kim Judge, Dania K. Saeed, Vanessa K. Wong et al., "Emergence of an Extensively Drug-Resistant *Salmonella enterica* Serovar Typhi Clone Harboring

a Promiscuous Plasmid Encoding Resistance to Fluoroquinolones and Third-Generation Cephalosporins." *MBio* 9, no. 1 (2018): e00105–18.

5   The news came out in *New York Times, Science, Washington Post, Telegraph*, and the *Economist*, to name a few.

6   World Health Organization, "Typhoid Fever—Islamic Republic of Pakistan," December 27, 2018, https://www.who.int/csr/don/27-december-2018-typhoid -pakistan/en/.

CHAPTER 26: TOO MUCH OR TOO LITTLE?

1   Jeremy D. Keenan, Robin L. Bailey, Sheila K. West, Ahmed M. Arzika, John Hart, Jerusha Weaver, Khumbo Kalua et al., "Azithromycin to Reduce Childhood Mortality in Sub-Saharan Africa," *New England Journal of Medicine* 378, no. 17 (2018): 1583–92.

2   Donald J. McNeil, "Infant Deaths Fall Sharply in Africa with Routine Antibiotics," *New York Times*, April 25, 2018.

3   Susan Brink, "Giving Antibiotics to Healthy Kids in Poor Countries: Good Idea or Bad Idea?," NPR, April 25, 2018.

4   Based on the author's interview with Dr. Thomas Lietman, September 5, 2018.

5   Shannen K. Allen and Richard D. Semba, "The Trachoma 'Menace' in the United States, 1897–1960," *Survey of Ophthalmology* 47, no. 5 (2002): 500–509.

6   Julius Schachter, Shila K. West, David Mabey, Chandler R. Dawson, Linda Bobo, Robin Bailey, Susan Vitale et al., "Azithromycin in Control of Trachoma," *Lancet* 354, no. 9179 (1999): 630–35.

7   Travis C. Porco, Teshome Gebre, Berhan Ayele, Jenafir House, Jeremy Keenan, Zhaoxia Zhou, Kevin Cyrus Hong et al., "Effect of Mass Distribution of Azithromycin for Trachoma Control on Overall Mortality in Ethiopian Children: A Randomized Trial," *Journal of the American Medical Association* 302, no. 9 (2009): 962–68.

8   Donald J. McNeil, "Infant Deaths Fall Sharply in Africa with Routine Antibiotics."

CHAPTER 27: VISA NOT REQUIRED

1   Based on the author's interview with Dr. Tanvir Rahman, February 2, 2019.

2   Ishan Tharoor, "The Story of One of Cold War's Greatest Unsolved Mysteries," *Washington Post*, December 30, 2014.

3   Based on the author's interview with Otto Cars, April 25, 2019.

4   For more information, see https://www.reactgroup.org.

5   Sigvard Mölstad, Mats Erntell, Håkan Hanberger, Eva Melander, Christer Norman, Gunilla Skoog, C. Stålsby Lundborg, Anders Söderström et al., "Sustained Reduction of Antibiotic Use and Low Bacterial Resistance: 10-Year Follow-up of the Swedish Strama Programme," *Lancet Infectious Diseases* 8, no. 2 (2008): 125–32.

CHAPTER 28: THE DRY PIPELINE

1   Michael Erman, "Allergan to Sell Women's Health, Infectious Disease Units," Reuters, May 30, 2018.

2   Sean Farrell, "AstraZeneca to Sell Antibiotics Branch to Pfizer," *Guardian*, August 24, 2016.

3   Pew Charitable Trusts, "A Scientific Roadmap for Antibiotic Discovery,"
    June 2016.
4   Muhammad H. Zaman and Katie Clifford, "The Dry Pipeline: Overcoming
    Challenges in Antibiotics Discovery and Availability," *Aspen Health Strat-
    egy Group Papers*, 2019.
5   Ibid.
6   Ibid.
7   Asher Mullard, "Achaogen Bankruptcy Highlights Antibacterial Develop-
    ment Woes," *Nature Review Drug Discovery* (2019): 411.

CHAPTER 29: NEW WAYS TO DO OLD BUSINESS
1   "How South Africa, the Nation Hardest Hit by HIV, Plans to 'End AIDS,'"
    PBS NewsHour, July 22, 2016.
2   For details on challenges associated with HIV drug access, see Michael
    Merson and Stephen Inrig, *The AIDS Pandemic: Searching for a Global Re-
    sponse* (New York: Springer, 2018).
3   Mandisa Mbali, "The Treatment Action Campaign and the History of
    Rights-Based, Patient-Driven HIV/AIDS Activism in South Africa," *Democ-
    ratising Development: The Politics of Socio-economic Rights in South Africa*
    (2005): 213–43.
4   Muhammad Hamid Zaman and Tarun Khanna, "Cost of Quality at Cipla
    Ltd., 1935–2016," *Business History Review* (2019).
5   Ibid.
6   Mandisa Mbali, "The Treatment Action Campaign and the History of
    Rights-Based, Patient-Driven HIV/AIDS Activism in South Africa," 213–43.
7   Kevin Outterson, "Pharmaceutical Arbitrage: Balancing Access and Inno-
    vation in International Prescription Drug Markets," *Yale Journal of Health
    Policy, Law, and Ethics* 5 (2005): 193.
8   Ibid.
9   Based on the author's interview with Dr. Kevin Outterson, June 12, 2018.
10 President Obama's executive order was issued on September 14, 2018, avail-
    able at https://obamawhitehouse.archives.gov/the-press-office/2014/09/18
    /executive-order-combating-antibiotic-resistant-bacteria.
11 Based on the author's interview with Dr. Tony Fauci, January 4, 2018.
12 William Rosen, *Miracle Cure: The Creation of Antibiotics and the Birth of
    Modern Medicine* (New York: Penguin, 2017), 122.
13 Helen Branswell, "With Billions in the Bank, a 'Visionary' Doctor Tries to
    Change the World," Stat News, May 6, 2016.
14 For more information about CARB-X, see https://carb-x.org.

CHAPTER 30: A THREE-HUNDRED-YEAR-OLD IDEA
1   M. Diane Burton and Tom Nicholas, "Prizes, Patents and the Search for
    Longitude," *Explorations in Economic History* 64 (2017): 21–36.
2   Ibid.

CHAPTER 31: SPOONFUL OF SUGAR
1   Ian Tucker, *Guardian*, May 21, 2011.
2   Based on the author's interview with Dr. James Collins, May 3, 2019.

3   Kyle R. Allison, Mark P. Brynildsen, and James J. Collins, "Metabolite-Enabled Eradication of Bacterial Persisters by Aminoglycosides," *Nature* 473, no. 7346 (2011): 216.

CHAPTER 32: CONFLICT INSIDE THE CELLS

1   Based on the author's interview with Dr. Houra Merrikh, February 14, 2019.
2   Jeffrey Roberts and Joo-Seop Park, "Mfd, the Bacterial Transcription Repair Coupling Factor: Translocation, Repair and Termination," *Current Opinion in Microbiology* 7, no. 2 (2004): 120–25.
3   Samuel Million-Weaver, Ariana N. Samadpour, Daniela A. Moreno-Habel, Patrick Nugent, Mitchell J. Brittnacher, Eli Weiss, Hillary S. Hayden, Samuel I. Miller, Ivan Liachko, and Houra Merrikh, "An Underlying Mechanism for the Increased Mutagenesis of Lagging-Strand Genes in *Bacillus subtilis*," *Proceedings of the National Academy of Sciences* 112, no. 10 (2015): E1096–105.
4   Mark N. Ragheb, Maureen K. Thomason, Chris Hsu, Patrick Nugent, John Gage, Ariana N. Samadpour, Ankunda Kariisa et al., "Inhibiting the Evolution of Antibiotic Resistance," *Molecular Cell* 73, no. 1 (2019): 157–65.

CHAPTER 33: SECURITY OR SERVICE?

1   Dr. Joanne Liu's speech to Barnard College, where she describes her background and upbringing, is available at https://barnard.edu/commencement/archives/commencement-2017/joanne-liu-remarks-delivered.
2   Based on the author's interview with Dr. Liu, August 10, 2018.

CHAPTER 34: ONE WORLD, ONE HEALTH

1   Based on the author's interview with Dr. Steve Osofsky, April 19, 2019.
2   Based on the author's interview with Dr. William Karesh, April 22, 2019.
3   Conference agenda available at http://www.oneworldonehealth.org/sept2004/owoh_sept04.html.
4   Yi-Yun Liu, Yang Wang, Timothy R. Walsh, Ling-Xian Yi, Rong Zhang, James Spencer, Yohei Doi et al., "Emergence of Plasmid-Mediated Colistin Resistance Mechanism MCR-1 in Animals and Human Beings in China: A Microbiological and Molecular BiologicalStudy," *Lancet Infectious Diseases* 16, no. 2 (2016): 161–68.

CHAPTER 35: BANKERS, DOCTORS, AND DIPLOMATS

1   Based on the author's interview with Dame Sally Davies, September 21, 2018. Additional biographical information available at https://www.whatisbiotechnology.org/index.php/people/summary/Davies.
2   Multiple British newspapers reported the interview on July 1, 2014. For example, see the report by the *Telegraph*, https://www.telegraph.co.uk/news/health/10939664/Superbugs-could-cast-the-world-back-into-the-dark-ages-David-Cameron-says.html.
3   Based on the author's interview with Lord Jim O'Neill, March 1, 2019.
4   Jim O'Neill, "Building Better Global Economic BRICs" (November 2001).
5   Final report of the O'Neill commission, along with other materials and infographics, are available at https://amr-review.org/Publications.html.

6   For details, see amr-review.org.

7   Marlieke E. A. de Kraker, Andrew J. Stewardson, and Stephan Harbarth, "Will 10 Million People Die a Year Due to Antimicrobial Resistance by 2050?," *PLoS Medicine* 13, no. 11 (2016): e1002184.

EPILOGUE

1   Marc Lipsitch and George R. Siber, "How Can Vaccines Contribute to Solving the Antimicrobial Resistance Problem?," *MBio* 7, no. 3 (2016): e00428–16.

2   Sara Reardon, "Phage Therapy Gets Revitalized," *Nature News* 510, no. 7503 (2014): 15.

3   Matthew J. Renwick, David M. Brogan, and Elias Mossialos, "A Systematic Review and Critical Assessment of Incentive Strategies for Discovery and Development of Novel Antibiotics," *Journal of Antibiotics* 69, no. 2 (2016): 73.

4   More information available at https://www.pewtrusts.org/en/projects /antibiotic-resistance-project.

5   See https://www.thebureauinvestigates.com/projects/superbugs.

INDEX

ABOUT THE AUTHOR

Muhammad H. Zaman, PhD, is a Howard Hughes Medical Institute Professor of Biomedical Engineering and International Health at Boston University. His work has been published in *Nature*, *Science*, and *Lancet Planetary Health*, among other magazines. In addition, his opinion pieces and columns have appeared in leading newspapers around the world, including the *New York Times*, the *Huffington Post*, *U.S. News & World Report*, *El País*, and *Japan Times*; on Al Jazeera; at the World Economic Forum; and through dozens of other outlets. He lives with his family in the greater Boston area.